In the Shadow of the Sabertooth

Also by Doug Peacock

In the Presence of Grizzlies:
the Ancient Bond Between
Men and Bears (with Andrea Peacock)

Walking It Off: a Veteran's Chronicle of War and Wilderness

Baja!

Grizzly Years: In Search of the American Wilderness

First published by
CounterPunch
and AK Press 2013

CounterPunch
PO Box 228
Petrolia, California
95558

AK Press
674-A 23rd St
Oakland, California
94612-1163

ISBN 978-1-84935-140-9

A catalog record for this book is available
from the Library of Congress.

Library of Congress Control Number:
2013930247

Design and typography by
Tiffany Wardle de Sousa.

Illustrations by Becky Grant.
(on pages: 25, 29, 32, 62, and 79)

Typeset in Minion Pro, designed by
Robert Slimbach for Adobe Systems Inc.
and Source Sans Pro, designed by
Paul Hunt for Adobe Systems Inc.

Printed and bound in the United States.

In the Shadow of the Sabertooth

A Renegade Naturalist Considers
Global Warming, the First Americans
and the Terrible Beasts of the Pleistocene

Doug Peacock

Author of *Grizzly Years*

Illustrations by Becky Grant

Table of Contents

CHAPTER 1 — **Repatriation and the Greatest Adventure** — 1

Sidebar: a note on dating and carbon-14 — 23

CHAPTER 2 — **The Lair of the Short-faced Bear** — 25
Forbidding Glaciers, Man-eating Predators and Poisonous Plants in Ice-age America

CHAPTER 3 — **Archaeology and the Shape of the Journey** — 63

CHAPTER 4 — **Invisible People: Life Before the Pre-Last Glacial Maximum** — 81
Ice-age People in the Far North of Siberia and America

CHAPTER 5 — **Mingled Fates of *Homo Sapiens* and *Ursus Arctos Horribilis*** — 105
Grizzly Bears as Proxy for Early Human Occupation of the Americas

CHAPTER 6 — **Braving the Northwest Coast During the Time of Icebergs** — 119
Maritime Learning and Innovation in North America

CHAPTER 7 — **Pre-Clovis People** — 147
The Significance of People in the Contiguous American States before Clovis

CHAPTER 8 — **Clovis** — 165
The Great American Invention?

CHAPTER 9 — **Endgame** — 185
Late Pleistocene Extinction and the Sudden Sunset of Clovis

Acknowledgements — 205

Bibliography — 207

Index — 215

For Andrea

CHAPTER 1

Repatriation and the Greatest Adventure

FOR THE PAST 12,000 YEARS, the world has enjoyed a relatively stable climate. Now, the time of predictable global weather has ended. The future will be unsettled, probably fiery and likely terrifying. Forces have been unleashed that threaten the future of our children. The early consequences of global warming have already settled over much of the planet.

Has this kind of climate change ever challenged humans before? It certainly has: Most recently right here in North America, when people first colonized this continent. About 15,000 years ago, the weather began to warm, melting the huge glaciers of the Late Pleistocene that calved off as icebergs and caused the oceans to rise. The Americas were probably uninhabited by people then but teaming with gigantic and fierce animals, many capable of killing and eating human beings. In this brand new landscape, the largest of all unoccupied wilderness regions humans would ever explore on earth, people somehow adapted to unfamiliar habitats and dangerous creatures in the midst of a wildly fluctuating climate. And they made it through. Along the trail of the first migration into the Americas lie challenging illustrations of courage and caution for modern people.

Though the rough outline of this journey is delineated by modern science, what drew me into the wild heart of the first Americans' story was the adventure, the exploration, the danger: Wondering what it was like to live with huge pack-hunting lions, sabertooth cats, dire wolves and gigantic short-faced bears, to hunt now extinct horses, camels and mammoth, to top a ridge somewhere in what would become Alaska and look out on unending wild country that encompassed two continents uninhabited by humans. It was the first and only time since Adam and Eve emerged from Eden that our species would come into so vast a land, a wilderness five times the area of Australia and never before glimpsed by an upright primate. Here lived beasts both fierce and wonderful. Some species had not previously encountered people, including a number of predators.

I can't think of a richer, wilder, more perilous time to live. For a person like myself who loves wilderness, this time in America had to be the ultimate journey—heroic, bold adventurers facing down danger at every bend of the river, surviving against impossible odds.

No doubt, these early Americans had no such picture of themselves; just getting through the day alive, finding adequate food and shelter in fluctuating habitats, embraced all shades of courage.

The inexorable force overshadowing all human migrations and accommodations to new environments at the end of the Pleistocene was climate change: The melting world of the Late Pleistocene opened the Americas to human colonization and contributed to the extinction of the big animals we call megafauna. The earliest Americans were ice-age pioneers who took advantage of the time of glaciers (seas were at a lower level) to cross over the Bering Strait. As the weather warmed about 15,000 years ago, they began to wind their way south into the contingent states. They no doubt boated down the Pacific coast, dangerously dodging icebergs in their small skin boats. Later, they found passages down through the glaciers, corridors between the two great American ice sheets. The great hunters, the Clovis people of mammoth fame, probably used the last ice-free corridor to sweep down south and, in a heart-beat of geology (a few hundred years), colonize nearly all of North America. They are often credited with hunting the huge beasts of the Pleistocene, the megafauna, into extinction. Without climate change, the Clovis invasion might not have happened and some of the megafauna might still be around. Of course, great controversy surrounds these assertions.

Today's shifting weather patterns, the shorthand we call "global warming," will far exceed anything our ancestors faced during the climate change 13,000 to 15,000 years ago. Yet, at the very end of the Pleistocene we find a major extinction event; 35 genera of mostly large animals suddenly disappear from the earth. The lethal combination of human activity and climate change are the chief suspects. Today, we are experiencing what authorities call the 6th Great Extinction—a much deadlier event than that of the Pleistocene—a crisis most believe to be driven by human-induced climate warming.

The two eras were of course quite different worlds. Thirteen to fifteen thousand years ago, the Americas consisted of two very sparsely inhabited continents; today's planet is packed with seven billion humans. The strategy for adaptation and survival during the warming climate of the ice age involved bold migrations impossible in the 21st century; direct comparisons of the two periods of climate change are often tenuous.

Still, I was curious and much of that curiosity was bundled around how humans perceive risk. Could there be a bridge between recognizing today's extreme climatic dangers and the Pleistocene lion crouched in the bush, waiting for two-legged ice-age prey? We evolved to deal with the predator. In comparison, present day "global warming" seems distant, harmlessly incremental or something that happens to remote strangers. For those ancient adventurers, however, the sabertooth was right there, every day—a pragmatic consciousness of great modern value.

We are left to imagine the details of everyday Late Pleistocene life: Archaeology and paleontology provide broad parameters of time and place but there is great mystery and much controversy surrounding the dates and routes by which humans reached the unglaciated core of North America. The hard evidence delineating the peopling of the Americas 15,000 years ago is sketchy. But what a vibrant life it must have been, lived in that wild ice-age topography whose considerable remnants are still with us today.

This book will attempt to dig into the last several millennia of the ice-age, that period of American archaeology for which there is yet the least scientific documentation.

It was, and remains, an incredible adventure—the wildest ride.

∾

How did this story take root in my own life? Like countless others, my early days were painted with archaeology's unforgettable colors. When I was nine, I looked for arrowheads in the muddy furrows of spring-plowed fields of Michigan. The map of my world hugged the banks and terraces of great rivers, guiding me along the low snaking ridges of ancient beaches down into cattail swamps where legions of Canada geese and whistler swans darkened the evening skies—the downy wildebeests of my watery Serengeti. In the blowouts on the sandy ridges, a profusion of fire-broken rock and brown chert flakes blanketed scattered arrowheads of another kind. Summer brought clouds of mosquitoes off the marshes and, at the edge of the swamp, the receding river revealed a pile of huge rough flint blades: A cache of material awaiting refinement into finished arrowheads. Come autumn, at age fourteen, I walked the ridges under the blazing maples and elms. A gust of wind skittered the leaves across a large anthill, a normal feature in this porous soil. What was different about the hill was the color of the sand: It was bright red with faint streaks of green. I would learn what this meant: Underneath, a stillborn child, consecrated with sacred red ocher, lay buried accompanied by Lake Superior copper grave offerings and a hundred triangular arrowheads.

The arrowheads shaped a central mystery and a lost way of life. They spoke of another world, an older more compelling world I wanted somehow to become a part of.

I skipped my way through awkward adolescence wandering my wilderness of marshes. Sometimes I carried an ancient Damascus-twist double barrel shotgun into the maze of channels. Pintail ducks exploded from potholes, startling my inattention. I pushed through the line of bulrushes walling off the river. The late summer breeze lifted off filaments of cottony fluff from the seed spike; I was reminded of the white flower pattern on a girl's pink underpants and I missed a brace of mallards rising off into the cirrus blue sky. The September sun sparkled on the muddy rivulets and, further upstream, glinted off flakes of chert and flint eroding from the riverbank. I splashed up the river's edge and came to a profusion of rocks, potsherds and flaked artifacts in the shallow water or tumbling out of the riverbank—three or four thousand years of prehistory. Among these lay four of the most perfect arrowheads I had ever seen: big, deeply corner-notched chocolate-brown chert projectile points, the largest over four inches long. No doubt, the big brown arrowheads had come from another cache or even a burial weathering out of the nearby bank. I would return but not to collect artifacts or look for the cache.

That next year I gave up my entire collection of arrowheads. I started reburying arrowheads and repatriating the ones I had kept as a boy.

Except for one. Later, I carried the largest of those chocolate-brown chert arrowheads into war. It protected me from countless enemy bullets and would prefigure the decades to come.

The arrowheads told a story but I didn't know what it was about. Our family had a trout fishing cabin on the upper Pine River. My grandpa and uncle built it out of wood scraps and tarpaper after the Depression. The cabin was where the stories were told. Grandfather narrated sagas of a gigantic brown trout hooked three times over a decade but never landed. He had a Chippewa Indian friend he sometimes ice-fished with. But the legend of the arrowheads was one story my grandfather never told me.

My father hauled me around to local archaeological meetings— amateur groups who would bring in a professional for a lecture—and helped me find books at the library. My dad made up wonderful stories about an Indian boy like myself, that he would write out over the years and mail to me from his distant Boy Scout postings, or, occasionally sneak into my bedroom (after all, I was a teenager) and ease me towards smiling sleep with his soothing woodsy tales.

I plugged away, tracking the trails of those ancient hunters, especially the earliest ones. Plunging into my backyard wilderness, I prowled those swamps and wastelands. The songs of warblers and larks ushered my forays into dark woodlands. Dusk suggested jeopardy. My child's universe of adventure edged into a larger world and I slowly began to crave wildness beyond the hills and cornfields. I know now that those fens, sand ridges and feathered herds flying at sunset gave rise to my own idea of home, one that had everything to do with discovery and a sense of the importance of wild exploration that eventually propelled a lifetime aimed at boundless horizons. Much of that value emerged from a child looking for arrowheads and then thinking about the lives of the vanished people who had made them. Where'd they come from and how did they live?

By this time, age fifteen, I had figured out that those sand ridges at 605 feet of elevation represented a post-glacial beach of the Great Lakes, lived on by Late Archaic people about 4,000 years ago. I found another red colored anthill and immediately called the anthropology department at the University of Michigan and talked to James B. Griffin, a giant, I later learned, in the archaeology of the eastern United States.

Griffin sent out two doctorial graduate students, Louis Binford and Mark Papworth. I tagged along on many a field reconnaissance and eavesdropped on conversations too sophisticated for my provincial upbringing. Papworth, especially, took me under his care; we slogged through muddy cornfields and paddled canoes down roily rivers looking for sites. Mark pointed out stands of wild marihuana and passed me a beer—knowledge and rites I had barely imagined. Some time passed, the University of Michigan got a grant for archaeological fieldwork in the Saginaw Basin and I was hired in 1960 as a research assistant on a dig of a site I had discovered when I was sixteen. Later, I attended the University of Michigan and took archaeology courses taught by these great men, learning about the peopling of the Americas and the bold hunters, called "Clovis," who once stalked mammoth during the time of the gigantic American beasts (now extinct) at the end of the Pleistocene.

But as a student, I was restless, aching for the Rocky Mountains and would quit alternate semesters to go West and pound nails for a living. All the time, the draft board was close behind.

In the spring of 1963, I was working as a core-logging geologist for a copper mine in southern Arizona. My U of M advisor, paleontologist John (Jack) Dorr, called and asked me to accompany him to the Alaskan backcountry on a three-month expedition to look for non-marine vertebrate fossils from the Tertiary era—an effort to correlate the Bering Strait land migration route theory for extinct horses and camels. We went everywhere: All over Alaska, the Yukon drainage, the Mackenzie River basin and the North Slope before big oil got there. The trip was a total academic and scientific failure. We found no such fossils, not a single one. It was one of the best times of my life.

We camped out on a braided river halfway between the Brooks Range and the Arctic Ocean. The bush plane that had landed on a gravel bar and dropped us off would be a couple weeks late in picking us up because of poor weather. We ran short of food, and worse, Jack was out of tobacco. Every day, we scanned the gray horizon for breaks in the weather: The plane never came. Jack smoked coffee grounds rolled in newspaper while I foraged the flats and hills for berries, fish and meat. We guilelessly wondered if we might end up wintering in this land of tundra and muskeg—we'd have to live like Indians or Eskimos. I fashioned a hook with feathers, made a fly, tied it with a leader to a short branch of dwarf willow, let the wind blow it over the sloughs and jerked countless grayling up onto the bank. Dr. Jack, who had collected sample skulls of nearly all North American mammals, asked me if I might bag an Arctic ground squirrel (whose skull was missing from his collection) for him with my pistol. I stalked the dry ridges, dodging dive-bombing gyrfalcons that nested on the low summits, and watched a coal-black wolf nearly my own weight disappear into the fog. That night I fried up the headless ground squirrel in bacon grease and wondered how the ancients survived in such a place.

I didn't get back to the North Country for a number of years. In 1966, I wrapped up the big chocolate brown arrowhead in a small roadmap of the northern Rocky Mountains and headed to Southeast Asia. The arrowhead kept me alive during firefights, grenade lobs, mortar attacks and friendly fire from helicopters and stray bombs. The map showed me what I wanted to stay alive for. After serving two tours as a Special Forces medic in the Central Highlands of Vietnam, I was finally repatriated to the Rocky Mountains, the Pacific coast of Alaska and British Columbia, the Arctic tundra, and the great deserts of the Southwest U.S. and Northwestern Mexico—North America, the land I loved the most. The maimed vet crawled back into the brush and lived with grizzly bears until his wound began to staunch, then struck out again walking the high country. The wild habitats of the West that represented my homeland were also the terrain of that great adventure—when the first people reached the western shore of the Pacific and found their way south. How would it have felt to be the first human to explore this uninhabited wilderness when huge lions, sabertooth cats and gigantic bears patrolled the land? I walked the wild ridges with these scenes in the back of my mind.

For four more decades I stalked these places and routes, following grizzly bears, often retracing the paths of prehistoric people, never with artifact collecting on my mind but rather with a sense of wonder and curiosity about how people might have lived in such habitat. Accordingly, I lived off the land, often alone, for weeks at a time in remote deserts, mountains, coastal British Columbia, including the Queen Charlotte and Goose Islands, a coastal route that was not completely glaciated during the last Ice Age. I walked point on a polar bear expedition in eastern Beringia (Beringia during the Late Pleistocene was the vast Arctic region encompassing the Bering land bridge, west from the Russian Far East, Siberia and much of Alaska, east all the way past the Mackenzie River in Canada), tracked Siberian tigers in western Beringia, stalked grizzlies throughout Alaska and all the way down into Mexico, slowly paddled down the Porcupine River where I found a mammoth tusk sticking out of the bank of a side channel and roamed the region of the ice-free corridor, ranging from the waters of the Yukon and Mackenzie Rivers south to the Rocky Mountain Front. For seven months a year, over fifteen years, I lived with wild grizzlies in the high mountains of the American West. I'm still here.

Archaeologists had enriched my life. After the war, I lived in the home of eminent anthropologist Edward H. Spicer, where I came to know Thomas Hinton and Bernard (Bunny) Fontana. For years, Tom and I camped up and down Sonora's Seri Coast, and Bunny remains a close friend. Henry Wright bailed me out of a jam one time. Mark Papworth read my first book, *Grizzly Years*, got hold of me through the publisher and we resumed our old friendship.

The past was close behind. My repatriation was already unconsciously tracking this great adventure story—the colonization of ice-age America by humans. A couple developments narrowed my focus on this tale into the brilliant sunlight of a Montana summer and the writing of this book. The first of these concerned an archaeological site.

∼

Back in 1968, just north of my house on the Yellowstone River, workers unearthed about 110 stone and bone artifacts that accompanied a child burial. The funeral offerings were consecrated—like those on the ridges of my youth—with sacred red ochre, an ancient burial practice that goes back nearly 100,000 years in the Old World. These grave offerings constitute the largest and most spectacular collection of Clovis tools ever found (the Clovis culture dates from about 13,100 to 12,800 years ago and was once believed to represent the earliest Americans who presumably dashed down the ice-free corridor from Alaska along the Rocky Mountain Front into Montana.) The one-and-a-half-year-old child is the oldest skeleton ever found in the Americas and the only known Clovis burial.

Partly because construction workers had discovered the burial and because rural Montana was far from the scholarly centers of Pre-Columbian archaeology, professionals largely ignored this stunning find and the significance of the site was dismissed or discredited in the scientific and popular literature for decades. At this time, archaeology reentered my life. With two professional friends, I helped organize a re-excavation.

Notebooks, 1999

I first heard about the Clovis Skeleton on a cold November day in 1998. The Livingston, Montana Natural History Exhibit Hall was sponsoring a tour through Paradise Valley where I live, just north of Yellowstone National Park, and a gruff, bearded, 55-year-old archaeologist, outfitter, and guide named Larry Lahren conducted it. Our group explored ancient bison-kill sites along limestone cliff faces and examined red-ocher pictographs that marked the entrance to a canyon just south of town. In passing, Lahren happened to mention a site he had studied north of Livingston, on veterinarian Mel Anzick's ranch—a place that held special significance for him. Intrigued, I invited Lahren to join me at the Murray Hotel Lounge for a drink.

Lahren has a reputation that matches his imposing physical presence; he's built like a football player, thick and hard, with a bit of a middle-age belly that belies the strength and quickness he once used to sweep three drunken cowboys off a Livingston bar. My friend the poet Jim Harrison had warned me, half joking, that it was OK to have two beers with Lahren, but that I should leave before he finished the third. We were on number two when Lahren started getting fired up about the importance of the Anzick site.

"It produced the only Clovis skeleton—period!" Lahren exclaimed. "But nobody in the archaeology establishment wants to hear it. But I know it's true."

Given the archaeological importance of Clovis artifacts, it seemed amazing that the only Clovis burial assemblage in the world had been found just a few miles away, and yet remained uncelebrated and almost unknown outside the professional literature. As Lahren continued his remarkable tale, however, I realized that when it comes to the Anzick site, missed opportunities abound.

One morning in June 1968, two local construction workers drove a front-loader and a dump truck out to the base of the elephant-head bluff. Mel Anzick had given the men permission to dig up fill for the local high school, and after Ben Hargis filled a dump truck, Calvin Sarver drove the first load into town.

Hargis continued working. He began punching into the scree at the base of the cliff with the bucket of the front-loader, and as he backed away with a full load, something fell down into the bucket, catching his eye. Bright red powder cascaded down the cliff from the place the object had fallen. Sarver returned to find Hargis excited: He'd found a very old and impressive-looking flaked tool.

That evening after work, Sarver and Hargis returned with their wives to explore the cliffside. They began digging with their hands, and almost immediately a huge chert blade, stained red, fell out. It was flaked on both sides, the sort of tool called a biface. Then another, and another—one made of yellow chalcedony, the next of red jasper. Stacks of big bifaces and spearheads spilled down the slope. Mixed in with the artifacts were fragments of a small human skeleton covered with red ocher; all the stone implements and bone tools were stained with it too. "We were up to our armpits in that red stuff," Sarver recalled recently. Faye Hargis remembers that they took the tools home and tried to scrub them clean—a task that left the kitchen sink stained red for a week.

Lahren, then a graduate student in archaeology at Montana State University, in Bozeman, heard about the find and asked to see the points, expecting to see weapons from a buffalo kill site, the sort that are common in these parts. He got his first look at the collection in Sarver's kitchen. There was some small talk, Lahren said, and then Sarver and Hargis went out and returned carrying ten five-gallon buckets full of artifacts into the house.

"I was speechless," Lahren told me. "I thought I was going to have a heart attack." He realized the two men might have found important evidence that could help solve the mystery of the identity of the first Americans.

Lahren told Dee Taylor, a professor from the University of Montana, about the discovery, and after identifying the points as Clovis, Taylor presided over a two-week dig in the summer of 1968. But the enterprise was troubled from the start. "It is almost enough to make strong men weep," he wrote later. The amateur diggers had "succeeded in taking almost everything that was there 'in situ.'"

Taylor's dismissal of the Anzick site established the attitude that remained prevalent for the next two decades. Artifacts from the Anzick site appeared on the cover of *National Geographic* in 1979, but the site was only mentioned briefly in the accompanying story about early Americans.

As Lahren and I sat that evening in the Murray Lounge, as I listened to this strange archaeological saga, I realized how passionate Lahren still was about the Anzick site, and I found myself catching his fever—a new outbreak of the enthusiasm I've had for archaeology since I was a boy. I envied the people who lived in that valley 13,000 years ago. I couldn't help thinking that the supreme American adventure had been the first one. When the humans first reached our shores, America was the greatest unexplored frontier on earth. Lahren seemed to feel the same way I did, and he clearly had unfinished business out in the Shields Valley.

And so I wasn't really surprised about what happened next. The bar was getting noisier, but we sat silently for a few minutes, and then Lahren said, "I'd love to get back in there with a crew and dig this the right way."

Those were the magic words. I immediately thought of Papworth. Once when he was running low on cash, Mark sold me his beloved 12-gauge Ithaca LeFever shotgun.

But then we lost touch. Three decades had passed since we last spoke when out of the blue I received a letter from my old professor. He'd read my book, *Grizzly Years*, in which I described taking vengeance on a particularly nasty rural phone booth. "I bet that was my shotgun you used to shoot that phone booth," the letter said. Enclosed was a business card: "Mark Papworth, Ph.D., Chief Deputy Coroner, Thurston County, Special Deputy-Homicide, Thurston County Sheriff's Office. Member of the Faculty, Evergreen State College."

I wrote back, "Dear Dr. P.: I shot that sucker six times with great satisfaction using your shotgun." Our friendship resumed. In the last decade of his life, Papworth and I were family, sharing his home in Arizona in winter and mine in grizzly country when the snows melted.

I knew that Papworth would be tempted by Lahren's scheme, and indeed he was. He agreed to join the team, saying, "It will be a last great adventure for this old man." For the rest of the winter and into spring, the three of us talked and schemed and brainstormed. In May, we visited Mel and Helen Anzick at their home near Livingston and asked for permission to resume the work that had come to a halt in the 1970s. To our delight, they said yes.

Perhaps the Anzick site was going to get its due. If not, we would at least have a hell of a lot of fun.

July 1999, 8 a.m. a warm, sunny morning, and our 11-person crew— the 1999 Anzick Excavation Team—is crowded around the sandstone outcrop, sipping morning coffee from paper cups. The ground rules include "no poking around" (because this is a consecrated burial site) and no serious beer drinking until 5 p.m. We've planned two short digs for this summer, squeezed in between Lahren's paying job for a mining company as a contract archaeologist and Papworth's family obligations. Standing at the base of the bluff, we see that a giant bite has been taken out of the slope by previous digging. We will clear this area back to the cliff and down to the bedrock, revealing the original layers. Although Papworth believes the unexcavated areas may contain multiple burials, we will avoid digging in undisturbed dirt.

Lahren and Papworth will call the shots; the rest of the all-volunteer team, including a geologist, an anthropologist, myself, some students and friends of Lahren's, will do the grunt work, cleaning away rocks and debris and getting the site ready for further study.

And so we work, hauling rocks away in wheelbarrows and sifting sand through screens, making sure we don't miss anything.

On the evening of the first day, as the sun begins to cast a soft golden glow on the cliff face, we welcome an invited guest: The co-discoverer of the site, Calvin Sarver, now in his 60s, has just arrived from town to tell us about that bizarre day 30 years ago when the cliff seemed to rain Clovis artifacts. He walks to the cliff and points at a spot on the wall six feet higher and 15 feet east of the place where both Taylor and Lahren had previously dug.

"It was right here," Sarver says. "Just about this high."

Lahren is stunned. "You're sure about that?" Sarver seems certain, although he grants that it's been 30 years. This is invaluable information, which my son Colin records on video camera. I ask Calvin why he hadn't cleared up all the previous archaeological misunderstandings: "Nobody ever asked me," he answers.

Unlike Taylor, who died in 1991, Lahren now has a chance to set the record straight. "You know, I just assumed Taylor excavated the right place," he says. "I can't believe it. We just sifted through his leavings. Well, I guess we better re-do this grid."

The question of DNA testing on the bones comes up. One of the Anzicks' five children, their 33-year-old daughter Sarah, is better qualified than most to consider the ramifications of DNA testing: She has worked as a molecular biologist since 1994 at the cancer genetics branch of the National Institutes of Health's human genome project in Bethesda, Maryland.

"Because the results could shed light onto patterns of human migration," Sarah wrote to Lahren and me in September of 1999, "the results could have profound significance for the Native American community. The Native Americans have been intensely concerned about all genetic testing, so the [National Human Genome Research Institute] has been working very hard to build a bridge with this community. Given this, we have a moral obligation to communicate with the Native Americans and to be sensitive to their concerns regarding the genetic testing of the Anzick site remains."

Meanwhile, there is another uncertainty: In recent months, as dealers continue to offer substantial sums for the Clovis artifacts he owns, Mel Anzick has apparently developed a new ambivalence about the potential wealth the artifacts represent. "It's like finding oil on your place," he said last fall. I was afraid of this: In our attempt to establish credibility for the Clovis burial, we advertise the site and its artifacts to all kinds of greedy collectors and pushy scientists.

On the evening of our second-to-last day working at the Anzick site, after the others leave, I climb to the top of the sacred elephant head and breathe in the immense space under the vault of the Montana sky, the landscape wild and free since the Pleistocene. May it stay this way, I whisper. But deep down, I know the story of the bones is far from over.

What emerged from our re-excavation of the Clovis child burial site were a number of fundamental questions and some clues as to how to go about answering them.

And these questions were the huge unanswered mysteries surrounding human colonization of the Americas: who were the First Americans? Were there people in North America before Clovis or, much earlier, before the last advance of the great ice sheets? These are two separate questions. Where did they come from and when did they arrive? How did they get down to Montana from Alaska or Siberia? Did they come by land or coast? Could they have come from Europe? How did early arrivals to North America ever survive the terrifying array of Pleistocene predators? What was the origin of the Clovis point (the manufacture of the signature tool, a superbly flaked and fluted spear head, some consider a "revolutionary" lithic technique)? Did it come from, say, Europe, Asia or was it a unique American invention? Finally, how did Clovis technology spread so fast on a sparsely inhabited continent? Both Clovis and the last of American Pleistocene megafauna disappeared at the same time, just over 200 years after the child was buried. Did the Clovis people hunt the mammoth and other huge animals to extinction or did climate change or an asteroid cause their demise?

These are the questions I plan to explore in the next eight chapters.

My two archaeologist friends and I also wondered if there was something special about this site: In addition to the fact that the burial contained the largest cache of Clovis artifacts yet discovered and the only Clovis skeleton, the Montana child burial provided hints that this find could be one of the oldest—there are older, though challenged, dates—and a key to understanding both the migratory routes of the First Americans as well as the origins of Clovis projectile point. Of course, a few people lived south of the ice before Clovis showed up. We'd work it out. Clues came from the geography of the site, the kind of stone used for the Clovis blades and projectile points, and the antler foreshafts (the detachable rods armed with the projectile point).

In the decade to come, the interpretation of the Anzick site would constitute a key argument by prominent archaeologists in the great theoretical wars surrounding the peopling of the Americas. Some of these professional readings do not match up with the factual evidence of the site and these misinterpretations have goaded my provincial defense of this local treasure.

∾

The second development was a simple observation. Five autumns ago, behind my Montana house and far up the Absoroka Mountains, the forest turned red. So did the tops of all the other mountain ranges in and around nearby Yellowstone park. You could see it from the highways. The region's whitebark pine trees succumbed to an invasive pine beetle on a scale of death none of us thought we'd ever see. And it happened so fast—not in decades but just a few years—that it took both concerned citizens and scientists by surprise. The reason the trees died is because the winters warmed up during the last decade and the mountain pine beetle, already active in the lower lodgepole pine forest, moved up a life zone into the whitebark and killed the trees. Nature controls the beetle by freezing the larva—cold temperatures of minus 30 to 35 degrees Fahrenheit for about five days in winter, depending on the thickness of the tree bark. Incidentally, whitebark pine nuts are the most important grizzly food in the Yellowstone region. With whitebark pine nuts eliminated from grizzly bear diets—and this seems to be the case—grizzlies in this island ecosystem will be severely endangered. The bears could be on their way out.

Here is an issue close to my heart. I have always argued, not quite glibly, that the fates of humans and grizzly bears are mingled, confined to a common destiny shared in the same habitats. If brother bear was going, could we be far behind?

Not everyone lives at the foot of a mountain range whose high forests have already been blasted by the effects of climate warming. Elsewhere the consequences are less visible, more elusive. We sense big changes are coming but for now life is good. Yet the threat is real. The precise problem seems to be that modern humans have difficulty perceiving their own true long-term self-interests; we don't quite see the evolving threat to our survival as a civilization or a species. There's no Pleistocene lion lurking in the gulch. But beyond the false invulnerability of our clever technology and the insulation of our material comfort, here prowls the beast of our time.

∼

As a naturalist of sorts and an advocate for wildness, I try to make a difference through my work. The central issue of my generation is the human perpetrated wound we have inflicted against the life-support systems of the earth, whose collective injuries are increasingly visible today as climate change. Should humans push through another population bottleneck, we will drag down much of the wild earth and almost all the large animals with us. And that's the rub: not that it's unfair, which it is, but can people thrive without the habitats in which our human intelligence evolved, that gave rise to that bend of mind we call consciousness? *Homo sapiens* evolved in wilderness landscapes that are in part still with us; can we hope to endure when that homeland vanishes?

When I decided to write a book about people first coming to America during the Late Pleistocene, I had the importance of wilderness and modern global warming rattling around together in my brain. What do they all have in common? The biggest wild landscape ever glimpsed by *Homo sapiens* was at the moment people set foot on the Americas— two huge uninhabited continents. All the prehistoric action takes place in lands whose remnants today we call wilderness. And, I believe, the conservation of wild habitats will play a decisive role in our attempts to adapt to the current shifting climate.

The time period in which most of this book unfolds is about 13,000 to 15,000 years ago, a time, like today, of convulsive climate change. The end of the Pleistocene in North America was a time of rising temperatures, increase in the release of Arctic methane gases, melting glaciers, acidifying oceans rising hundreds of feet and massive extinctions. Could the earlier adventure, I wondered, inform the latter in any pragmatic fashion? Are there lessons in the story of early Americans adapting to a changing climate in an uninhabited human landscape prowled by huge cats and gigantic bears? The sudden emergence and disappearance of Clovis culture along with the extinction of North American megafauna are certainly related to changing weather patterns. While this ancient tale is not directly connected to 21st century global warming, the specter of climate change is the mammoth in the room.

∾

"The Greatest Adventure" is the story of the journey that launches human life on this continent, an epic trip capped by the near-synchronous appearance of the first major human occupation of the Americas and the extinction of the giant megafauna. I might add that this is where my story will end: After the twilight of Clovis and the sudden disappearance of America's great Pleistocene animals, the amount of anthropological data surges and a rich library of emerging archaeological material illustrates human life between the time of the Clovis demise right up to historical contact of Native Americans and European culture. Many books do a fine job of telling this story. Prior to Clovis and the demise of the big mammals, the archaeology is more controversial and there isn't much of it. This scant material is lobbied from partisan camps with arguments shouted out to the interested layperson who is encouraged to pick sides.

The Clovis colonization of the Americas climaxes around 13,000 years ago and ends less than two hundred years later with a sudden shift in climate and the final demise of the huge Late Pleistocene animals. These phenomena intersect in time and causation; we still don't know what induced the cycles of climate change, if human hunters brought down the megafauna or if, conversely, the fierce, huge predators of the Late Pleistocene impeded movement of people throughout the Americas. The precise timing of these events and the etiology of the collapse of American large mammals are yet cloaked in controversy.

Who were these ancient pioneers? They were modern humans, like us, with a different set of skills and priorities. These first Americans were true children of the ice whose ancestors had come up from the temperate lands of Asia, edging north and crossing the Arctic Circle 30,000–40,000 years ago. They were hunters, especially of big game and migratory birds. These men and women possessed the social cohesion of nomadic bands, spoke an unknown language and used a tool kit geared to survive and thrive in the frozen North Country. By at least 24,000 years ago, people were wearing—evidenced by carved ivory figures—fur clothing on the shores of Lake Baikal in southeast Siberia. During the long nights of winter, the elders told tales that reached back millennia, an oral history and collective memory that embraced geography, climate and a detailed knowledge of plants and animals, which provided a template for the rush of discoveries they would encounter in the New World.

Traveling into American is unique in the history of human expansion. The first Americans faced the largest unexplored frontier in the history of colonization, two huge continents with no trace of people. The ice-age landscape was a hunter's dream, teeming with huge animals, many never seen before by humans, including a dangerous array of giant carnivores.

The final migration, the firmly documented Clovis colonization of the Americas, took place at a rate unprecedented in the global archaeological record: Within a time period of just a few centuries, these Clovis people left their distinctive spear points from Montana to Florida, from New York to Central America.

(There are two later migrations of people into America: The Da Dene around nine thousand years ago and the Inuit-Aleuts a few thousand years ago. These arrivals are not covered in this book, which ends at the time of the great megafauna extinctions around 12,900 years ago.)

This book is the story of those human migrations into the Americas, beginning with the ephemeral ice-age people at the peak of the glaciation, to the bold mariners who no doubt traveled the northwest coast during the time of icebergs and finally with Clovis and the extinction of the megafauna.

$\sim$

Here arguably is the world's greatest adventure story: Ice-age hunters exploring a brand-new wilderness, braving raging rivers, crossing glaciers, encountering never-before-seen giant creatures—surviving in these shifting, uninhabited landscapes amid a rapidly changing climate. This journey is our unclaimed American Odyssey.

The notion of adventure includes risky undertakings, hazardous journeys with uncertain outcomes. The accounts of European exploration of the Polar Regions, Lewis and Clark moving up the Missouri River, John Wesley Powell going down the Colorado or surviving with Cabeza de

Vaca are saturated with adventure. Today, with fewer blank spots left on the map, true adventure is a more elusive attainment; we are often forced to design our modern adventures, complete with magazine, book and movie deals, vaguely hoping for unexpected turns and slight misfortune. Yet, even vicariously, we still need this adrenaline-fueled hope called adventure, crave it, love it when we emerge at the take-out, a changed person but alive and looking at the world anew.

Though this journey was traveled millennia before written history and is only faintly delineated by our modern science, the tale looms as an untapped reservoir of human inspiration, as useful to people today as the most epic stories ever told around the campfire or in our books and folklore. It's hard to imagine a more vital time to live than the Late Pleistocene in North America. Everything was new, the living dangerous, the daily routine utterly engaging.

This story stands in opposition to the history lesson I was taught in school—the pap of pilgrims conquering a dark and foreboding wilderness, of subduing godless savages with disease and blunderbusses, of Mayflower and Manifest Destiny. The Greatest Adventure begins at the opposite side of the continent, enters a land bountiful without parallel, the bright habitats beckoning with adventure, sizzling with life and devoid of any trace of human occupation. But it also bristles with dangerous beasts, formidable water crossings and massive ice fields. The Greatest Adventure was a much tougher trip through paradise.

The great American naturalist John Muir (presaging E. O. Wilson's "biophilia") believed his passion for nature came from a "natural inherited wildness in our blood." Muir believed that natural selection created that passion and that it was permanently buried in our brains and genes.

Our own organic consciousness evolved within wild habitats from the African savannah all the way to the frozen tundra of the North. Evolutional awareness was shaped by the mammoths we hunted, by the great cats and bears who sometimes stalked us. And, as the lynx still sculpts snowshoe hare evolution, what forces today yet hone the human mind that was born of foraging? Modern people sometimes insist they exist apart from nature, the conditions that gave rise to human awareness—the habitats whose remnants we now call "wilderness." But today nature has reasserted herself. The signs are dire. Will we heed the warnings? The Pleistocene predators are gone. A child in danger, a dark alley or a personal brush with tragedy generates an appropriate emotional response far more easily than the distant but predictable ocean rise that could displace a billion starving human strangers. Once again we live in dangerous times and navigating these treacherous waters will require sharpening that ancient perception of risk. It might not be a bad idea to try to hang on to some of that original landscape, like the wild Pacific coast or the cordilleras of the American West, habitat for survival, where utilitarian adventure still smolders.

$\sim$

Why not have the professional archaeologists tell this story? Good question. Those archaeologists who have written such books jealously guard their territorial prerogatives. Even when writing books for the general reader, archaeologists tout the unique value of having been inside the authenticity-bestowing room when credibility proclamations are awarded. Some insiders question whether salvage archaeologists are qualified to criticize academic papers and if non-scientific but mainstream magazines should be writing about archaeological issues. Maybe they are correct.

I had a few reservations about the field; my archaeologist friends instilled in me a healthy skepticism, particularly about the specialized study of First Americans; they acknowledged that setting up two hypotheses as if the truth of one negates the truth of another is a persistent problem of archaeological intellectual life and noted that sample sizes tend to be small and correlation does not explain causation. Nonetheless, tough thinkers surround the profession and I wondered if this dialogue could benefit from an outside interpretive voice.

Underneath, I may have also sensed a bit of resentment (primed by my Montana experience of mainstream archaeology's dismissal of the Anzick Clovis burial and scientists' subsequent shifty scramble for the child's skeletal remains) at the injustice that the larger story, the story of all our people—our American Creation Myth—was patrolled and constrained by an academia whose own literature was frequently composed with a territorial imperative. Yet the field encompassed the landscape of legends, quartered in our childhood dreams of fossil giants and arrowheads. Technical scholarship sometimes swallows the best of this adventure: Could a naturalist's take on the archaeological material liberate this tale?

I read up on the subject, consuming volumes of material trying to sate my curiosity with the lives of these ice-age pioneers; the depth and range of the controversies surrounding this field of early American archaeology, along with the vehemence and niggling with which these academic wars are waged, astounded me, fascinated me. The richness of the material leapt off the driest scholarly page. Although many strive for the juice of modern discovery, the dozen or so most recent books on the topic, seemed insular (written by archaeologists with other archaeologists in mind) and devoid of the older vitality that lies at the heart of this incredibly journey.

Paradoxically, this is a tough tale to be told by an insider. Given the paltry stack of hard evidence, there's too much academic territory at stake.

≈

In 2007, I applied for and received a Guggenheim Fellowship. I spent the next two years reading related scholarly papers; it's interdisciplinary—archaeology, paleontology, genetics, linguistics, glaciology—but finite and accessible to an informed researcher, such as myself with a dusty degree in geology, graduate study in anthropology, a background in archaeology and the natural sciences. I also talked with a few primary researchers in the field. I can see why this tale is hard to bring to life but, after another year of following stories in the press, and a 2011 Fellowship from the Lannan Foundation, I thought I'd give it a shot.

Why write a book about a journey that took place many thousands of years before any written record, especially during a time when the world seems on fire? After all, these days are the most dangerous times we have seen in the history of the earth. Beyond the agony of modern wars, disease, economics, genocide, torture and starvation, the planet itself and its support systems are in peril. We are experiencing the largest rates of plant and animal extinction on record, rivaling the massive extinctions of the Cretaceous that knocked off the dinosaurs. All around the globe, the air is poisoned and the oceans over-fished. Climate change threatens all species, including humans. Global warming is not a passing phenomenon. It will be there at the end of the day and at the end of our lives. Revisiting an ancient puzzle that unfolded 13,000 to 15,000 years ago might be a waste of time and energy. So, again, why go back and track the odyssey of these bold first Americans?

Human adaptation to climate change is the common underlying theme. However thin the threads of evidence illuminating adaptation to ice-age global warming, I wanted to follow them to some speculative conclusion, even if it merely adds up to a wild guess. Direct comparisons between the two periods of climate change are impossible, rendered unproductive by an unmistakable lack of hard evidence. But the older journey is a great adventure, closer to mythology than science. Life in the Pleistocene is our original emergence story. Our own American Odysseus was out there fighting off ice-age sabertooths and bears with his spear while pursuing mammoth and other giant beasts.

In this spirit, I thought a good adventure story, occasionally with constructive parallelism, might spur us to open our hearts to the undeniable truth that we are again devouring more of the earth's resources than she has to spare. The book is also a celebration of the North American continent: An exhilarating tale—less prophesy than parable—spun along the lines of exploration in a brand new world beset by the storms of change. The hub of this story reveals adaptation by people coping with extreme climate change, the driving force of our evolution. Tracing the movement of the first people of the Americas' is ultimately an optimistic trip full of fun and excitement—a message of hope and courage we all could embrace.

A NOTE ON DATING AND CARBON-14

Whenever possible, I avoid using radiocarbon dates in this book. Of course, carbon-14 dating is the bread and butter of archaeology, which is ultimately rooted not in the amazing technology of accelerator mass spectrometry but the care and accuracy with which the carbon sample and its provenance are reported. For an outsider to question a particular radiocarbon date might be tantamount to calling the scientist a liar. Archaeology has a colorful history of great big liars (from Piltdown to Sandia) but I leave it to other archaeologists to state reservations about radiocarbon (and other methods) dates from key locations.

Briefly, the isotope carbon-14 (C14) is absorbed by plants from the air and moves on into animals until the organism dies. C14 then slowly reverts to N14. Half of it is gone in 5,730 years; another half of that, 25%, decays after 11,460 years and so on. The usefulness of radiocarbon dating fades rapidly for objects older than 45,000 years old. Dates are stated in years before present (BP) with "present" defined as 1950. A range of potential error, plus or minus in years, is provided with each analyzed sample. Calibration of radiocarbon dates to calendar years is neither linear nor especially logical; variation of cosmic particle bombardment from the sun and relative amounts of CO_2 stored in the ocean or air all play hell with recalibration. Tree ring chronology provides about twelve thousand years of comparison.

I have arbitrarily tacked on a couple thousand years to radiocarbon dates for the period of 15,000 to 12,000 years ago in order to discuss the time of the Great Adventure: Thus the radiocarbon date, 11,000 C-14 yr BP+/- 75 becomes a rough 13,000 years ago, in this case less than a hundred years off the true recalibrated date. For the purposes of telling my story, that's generally close enough. On occasion, radiocarbon dates are necessary to pin down the precise timing of the opening of the ice-free corridor or the brief panoramic window of Clovis. When you see an exception to rounded-off dates in this book, such as 13,300 or 12,900 years ago, that means the data is translated from the radiocarbon calendar and has been cited several times in peer-reviewed literature. I'd like to attempt to keep it simple.

Here is a sample of some rounded off dates that appear several times in the text:

30,000 years ago: Marks about the earliest date humans could have appeared in North America. The evidence? Not much: A single site in the Siberian Arctic and inferences from genetic and linguistic studies on extant populations.

15,000 to 27,000 years ago: A very cold time of advancing glaciers. By about 20,000 years ago the ice sheets were at their maximum, closing off all routes from Alaska to lower North America. The great megafauna still roamed the ice-free far north but there is no record of humans in the Arctic during this period.

13,000 to 15,000 years ago: Very rough dates that denote the time of the so-called pre-Clovis people. It was a time of global warming and rising seas. The archaeological data indicates a couple sites around 14,000 years ago in northeastern Siberia and, about 700 years later, several more along the tributaries of the upper Yukon River in Alaska. South of the ice, a number of credible pre-Clovis dates come from the United States and South America.

12,800 to 13,100 years ago: This is the time of Clovis, which probably begins several hundred years earlier than 13,000 years ago, by which date the culture, marked by its iconic projectile point, was full blown and spread across the southern half of the continent. There's lots of archaeology to pin these dates down. About 12,800 years ago, the warming period is interrupted by a cold snap. The American megafauna, which might have been in decline for a few hundred years, suddenly goes extinct, and Clovis disappears from the archaeological record.

CHAPTER 2

The Lair of the Short-faced Bear

Forbidding Glaciers, Man-eating Predators
and Poisonous Plants in Ice-age America

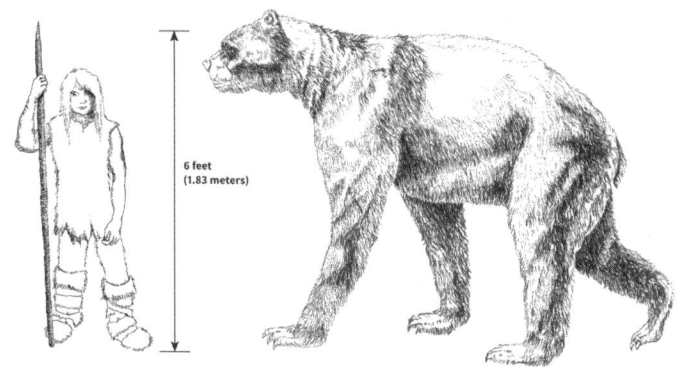

6 feet
(1.83 meters)

THE GREAT ICE FIELDS OF the last glacial advance loomed across the entire North American continent and at their maximum around 20,000 years ago constituted an impenetrable barrier to human migration. The glaciers themselves served as barometers to ice-age peoples and when they began to melt about 15,000 years ago they signaled the onset of approaching climate change.

Even today, I note that the question of where the ice is located still resonates in my personal life. In my small world of mountains and wilderness, recognition of the hard face of modern global warming comes in large measure not from reading the pile of scientific research which I receive most every day, but from old friends and some new ones who climb mountains. I am blessed to call among my friends a handful of the world's great mountain climbers. They climb in British Columbia, Alaska, the Andes, Africa and, most of all, in the Himalayas. My small forays and treks along the Continental Divide of the Northern Rockies and coastal British Columbia are meek in comparison to their expeditions to the high white wilderness of glaciers that mark the roof of the world. But the topic is the same: the disappearance of the world's glaciers, especially the finite ones we know well and explore like the body of a secret lover.

My own fragile and diminutive mistresses lie along a great traverse of grizzly bear country along the Continental Divide in Montana's Glacier National Park. The trek starts from the paved road the locals call "Going to the Six-Pack Highway" and departs from all trails as it leads you north towards Canada. It's not an easy bushwhack and the weather can shut you down. I've made the traverse over a half-dozen times beginning forty years ago, often alone or with a trusted friend and, the last, just a few years ago, an abbreviated trip with my own daughter.

As you climb up to the divide and wind around the peaks, Glacier's famous ice fields come into view. Most are tiny hanging glaciers clinging to the northeast-facing cliffs. A few of the larger ones have names: Vulture, Gyrfalcon or Two Ocean. You walk on snowfields and the tundra of the high country, where you can scarcely put a boot down without stepping on a grizzly dig. You finally scramble your way out of a perfect glacial cirque punctuated with a circular turquoise tarn, using your ice axe here and there for balance and backsliding, and step onto the Continental Divide. It opens up into a narrow expanse of subalpine splendor with stunted fir trees framing chains of shallow mountain lakes. Grizzly sign is everywhere; some of the high meadows look plowed by the long-clawed bears who dig corms and tubers from the thin soil.

Once I followed a well-used animal trail around the shore of small lake up there and almost stumbled into a garbage can-sized dish-shaped depression next to a tree; the grass at the edge of the grizzly daybed sprang upright from the weight of the bear who had just arose from his nap. Considerably more cautious, I crept down the trail. That's way too close.

In the 1970s, the small glaciers at the head of Valentine Creek still contained blue ice. By the late 1980s, these glaciers had become mere snowfields. Twenty years later, the snow was gone. Now, the ribbon of ice above lonely Gyrfalcon Lake is but a shadow of its hefty parent only thirty-five years ago. On my last traverse of the basin, I climbed up to the hanging glacier and broke off an icicle, sucking the water out of it like a last good kiss.

All this is but a brief moment in a single life watching small, beloved ice-fields shrink and die. A similar tragedy is playing out in the Wind River Range of Wyoming, where large glaciers, though diminished, still flow. My friends tell similar stories of the giant ice fields of the Himalayas. It is a tale of loss, of the end of something beautiful in a melting century.

One can imagine the minds of ice-age explorers dancing with images of glaciers. The first Americans of the Late Pleistocene would not have walked or boated down the icy defiles without a distinct foreboding of approaching change and an appreciation of the coming need for humans to pioneer new habitats, to confront these never-before-seen animals and endure shifting climates.

∼

The Pleistocene or Great Ice Age lasted about 2 million years (published estimates run from 1.6 to 2.6 million years), during which the northern polar caps surged and ebbed on a cycle of every 100,000 years or so. The last advance, the Wisconsin, started almost a hundred thousand years back in North America and, within that cycle, oscillations in the ice produced smaller advances and retreats. The Late Wisconsin, 12,000 to 38,000 years ago, was generally a time of expanding glaciers that reached their maximums about 20,000 years ago. This period is when humans first show up in North America.

Non-polar glaciers often have their origins in the high country when a cooling climate dumps more snow in the mountains in winter than melts during summer—accumulation exceeds ablation. As the snow builds up over decades and centuries, the pressure causes snow to granulate. The weight of ice compacts the glacier until, like a plastic, it begins to flow downhill. The rivers of mountain glaciers may coalesce and become ice sheets, like the Cordilleran of western North America.

The history of ice in North America is important because it outlines the great mysteries of continental colonization: Who were the first Americans, when and how did they get here and what routes did they take to get south of the ice? Since the archaeological record from the American Late Pleistocene is not robust, reconstruction of ancient environmental habitats contributes substantially to understanding when people could live there. And these habitats depended on where the glaciers were. The ice altered environments and climate. Constantly changing vegetative communities, chewed into a patchwork by the huge Pleistocene grazers and browsers, were the norm. The presence of the great ice sheets carved these plant communities into arrangements for which there are no modern equivalents.

Northern North America probably lay covered by ice during the Last Glacial Maximum (about 18,000 to 20,000 years ago). The Laurentide ice sheet buried the land from the North Atlantic coast to the Rocky Mountain trench, south into Ohio. The Cordilleran ice sheet came out of the Coastal Range and Rocky Mountains, huge valley glaciers coalescing into a composite sheet, smaller than the Laurentide, covering northwestern North America. Temperatures in Greenland during the last glacial maximum were 41 degrees Fahrenheit cooler than today. A key issue for archaeologists is whether there was a corridor between the two sheets sufficiently benign to permit human migrations coming south out of Beringia. Textbook maps often show the two great ice-sheets conjoined. Glacier erratic trains (boulders carried by the glacial ice) on the foothills of the Northern Rockies indicate this was the case 18,000 years ago. But there is conflicting evidence. The Ice Free Corridor (IFC), and other pre-last glacial maximum routes between the ice sheets, could have been open much of the time during the past 30,000 years. The debate continues.

People moved around the edges of these glaciers, looking for food and shelter from the cold. The Greatest Adventure began whenever those bands of Siberian hunters moved east across Beringia into present day Alaska. That crossing could have taken place anytime beginning about 30,000 years ago. The Bering Strait land bridge was almost certainly open 10,000-27,000 years back. Dating for the last stages of the Ice Age is imprecise; the geography of ice in North America, which areas might have been ice-free, is not clear. Many archaeologists believe the far north of Late Pleistocene Siberia was too damn cold for people to live there. Since no sites dating between 30,000 and about 14,000 years ago have been found in northeastern Siberia, some scientists think these children of the ice retreated south to sub-Arctic central Siberia during this time. Similar claims about the inhospitable climate—bleak, frigid, uninhabitable—are made for the same period in Alaska.

Approximate extent of ice during the Last Glacial Maximum.

But even ice-ages have summer time. Just because we haven't yet found an archaeological trace, that doesn't mean hunters were not eking out a living in these frigid climes. In fact, plenty of big game, the Pleistocene megafauna, roamed Beringia before and during the Last Glacial Maximum (LGM) when glaciers blocked all routes south; the Arctic, though colder before and during the LGM, was, botanists think, more grassy than shrubby offering an abundance of food for mammoth and other grazers who were in turn stalked through the snows by their gigantic predators.

The fossil record of Alaska and Canada confirms the presence of these animals. Now-extinct species of camel, long-horned bison, tapir, deer, antelope and horse ranged the tundra and grasslands. Great herds of caribou gnawed the northern lichen and bison grazed the open plains. Hidden in the draws and breaks were huge American lions, big dire wolves and gigantic short-faced bears. The deglaciated valleys were wet, the high benches speckled with pothole lakes, springs and ponds frequented by giant beaver. Mastodon browsed the edges of boreal forests; small groups of mammoth roamed the open country.

~

A spectacular array of very large animals lived in North America during the Late Pleistocene. Most of this astonishing menagerie of megafauna, along with some smaller genera of creatures, vanished suddenly nearly 13,000 years ago. The bulk of these animals were unusually large. Here's a sample bestiary:

Most iconic was the Columbian mammoth, monster of the plains, and its smaller cousin (they probably interbred) of the North, the woolly mammoth. Standing several feet taller than the largest elephant ever measured, these grazers no doubt traveled in matriarchal herds, like today's elephants. Massive, long tusks spiraled to a point and sometimes crossed. These creatures, some believe, were the spiritual and material center of Clovis culture and show up in a dozen kill or butchering sites. A small population of woolly mammoth survived on Alaska's Wrangell Island until about 4,000 years ago.

The American mastodon was also hunted by Clovis people, but perhaps not as frequently as mammoths. Shorter and stockier than the mammoth, this browser of spruce trees was a solitary animal often found in forests.

Clovis people also hunted a four-tusked cousin of the mastodon, the gomphothere, down in Sonora, Mexico about 13,000 years ago. This elephantine creature was found more commonly in South America and was, prior to the Sonoran discovery, believed to have gone extinct 30,000 years ago in the region.

Several kinds of giant ground sloth roamed the land and ranged in weight from about 200 to 6,000 pounds. The big ones had huge claws and one would think they would draw attention from anyone wandering the grasslands.

Early Americans no doubt hunted other American animals, for which we have no archaeological record of association. Chief among them would be horses, camels, tapir, peccary and, in the North, Saiga antelope.

A gigantic long-horned species of bison roamed the Pleistocene steppes and plains, along with herds of smaller buffalo. Bison, along with caribou and musk ox, important prey animals for Clovis as well as later human hunters, survived the great megafauna extinction around 12,900 years ago.

Smaller animals, but gigantic for their kind, included 350-pound beaver, armadillo and the glyptodont, a mammal almost ten feet long with armored shell, head and tail. Giant carrion birds teetered over the landscape, including condors with 20-foot wingspans. A deer called the stag-moose (slightly larger than a modern moose) displayed some of the biggest palmate antlers ever found on a mammal; these antlers are frequently preserved in fossil deposits. We don't know anything of the relationships of such creatures with humans but the animals must have painted the Pleistocene landscape with shimmering colors scarcely dreamed of today.

Preying on the grazing and browsing animals were giant carnivores. The biggest was the North American short-faced bear weighing in at over a ton. This long-legged giant could have been an omnivore but others think it lived by exclusive scavenging and predation. More on *Arctodus simus* later in this chapter.

The most effective American predator might have been the Pleistocene lion, same genus as today's African variety but eight-feet long with some of the biggest cat craniums ever measured, an animal that prowled North America and northwestern South America. The big brains, some suggest, indicate a highly social, pride-hunting predator.

The prototypical American Ice Age carnivore was the sabertooth cat, a stout, powerful predator with six-inch upper canines. The lovely name, *Smilodon fatalis*, says it all. This sabertooth was about the size of an African lion and is believed to have been a solitary ambusher of prey. Likewise, the scimitar cat had long, sharp serrated fangs perfect, they say, for slashing baby mammoths. As the second kind of sabertooth, the scimitar's teeth were nowhere near as long as Smilodon. The American cheetah, twice as big as the one in Africa, was also on the scene.

Wolves functioned much as they do today but probably scavenged more. The dire wolf and Beringian gray wolf had unusually heavy jaws and crushing teeth; the paleontological guess is that they both scavenged and hunted in packs. Dhole dogs, coyote and fox followed the flocks of condors, buzzards, crows and ravens to the kill sites of big cats and, eventually, ice-age Americans.

Los Angeles's La Brea tar pits hint at the spectrum of Pleistocene predators, unfortunate enough to mire in the tar. The Natural History Museum of Los Angeles lists these numbers: 4,000 dire wolves, 2,000 sabertooths (not scimitar), 80 lions, mostly male, and 30 short-faced bears. It's theorized that the sabertooths came in for mired bison and dire wolves cleaned up after them. Of course, tar pit death traps are not the same as kill sites: Carnivore behavior and selection (death) might be atypical at La Brea.

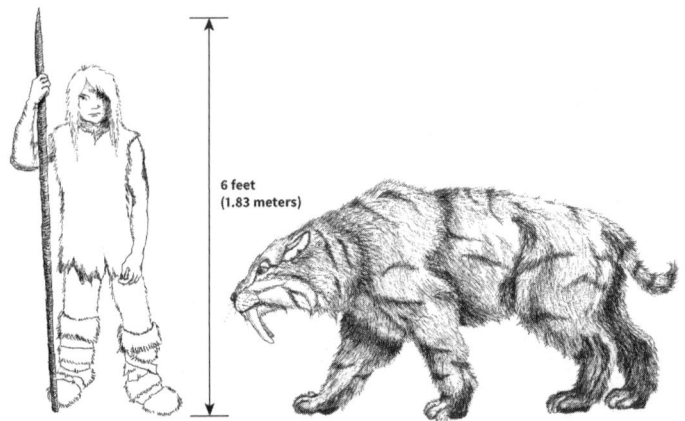

6 feet
(1.83 meters)

∾

As much as the ice sheets impeded human migratory routes, they also shaped the kind of people they would become. Surviving in the Arctic made the people tough; hunting big game and defending their kills against bears and lions could have infused hunters with edgy audacity and made them bold enough to go after the biggest game, at all costs. Eventually, some experts think, their experience north of the ice and in the ice-free corridors prepared them to hunt the American mammoth, the largest beast of all. This kind of prey required specialized lithic (stone-working) technology and may have spawned the iconic Clovis point. Much ink has been applied to papers debating these topics: Chapter 8 will add some more.

For the earliest American travelers, the ice must have seemed all encompassing, pervading their dreams and chiseling their angular features with blasts of Arctic wind. Coming down any route from Beringia, through icy defiles, the cold beauty of the glaciers would have dominated all landscapes. One could imagine the first Americans consciously aware of the growing, then shrinking glaciers, witnessing the melting white wilderness with a measure of alarm. Along the coast, the rising ocean would have haltingly inundated the forest, inch by inch. Climatic fluctuation, the retreat of the ice and the declining mass of the megafauna were not imperceptible changes to the first American explorers over the years. Beyond the daily struggle to survive, the wind was alive with the palpable scent of the regeneration of the earth—icy winds off the glacial front, the warm chinooks blowing up the Rocky Mountain Front and a wind bearing the fetor of dying beasts. How exciting, how terrifying a time to live: The last days of the Ice Age.

∾

The wide aim of this book is not so much to sort out the archaeological and other arguments (a great story) about people coming to America in the last days of the Pleistocene, but to inquire how people might have responded, bearing witness to radically changing environmental conditions. Though fundamentally unknowable, the question is worth some speculative consideration. Today, we approach a world we might not recognize by the end of this century. "Global warming"(often softened by the term climate change) is a catchword we can conveniently ignore with our modern technology and cultural insulation. Should our local weather warm up by a few degrees, who cares? But the extremes of global warming—widespread drought, floods, fierce storms, frigid winters in temperate zones and fiery heat—are the big enchiladas of global warming. These intense events can dramatically shift the limits of agriculture, create uninhabitable deserts the size of continents and break down the boundaries of what we call civilization. That this could happen within our lifetime does not seem to sharpen our perception of the threat. The climatic shifts of the Pleistocene might look quite mild in comparison to those of the 21st century.

What does it take to see the shadow of the sabertooth in the present day bush?

That particular conundrum is the challenge of this book.

∾

While writing a book about the past, our own crisis of climate runs through my head. Every day. I find myself grasping for comparisons that aren't quite clear. I have a friend, old-fashioned in his communication technology, who sends me news clips from newspapers and magazines by mail. He knows I'm working on a book about the Pleistocene and he wants to keep my mind straight. To keep my nose to the grindstone, I post his worst scenario for today's global warming disaster to the back of a map of Yellowstone Park, where I can't avoid looking at it.

> "By 2100, the Earth's population will be culled from today's 6.6 billion to as few as 500 million;" says James Lovelock, independent scientist and father of the Gaia theory, "billions of us will die and the few breeding pairs of people that survive will be in the Arctic where the climate remains tolerable." Lovelock also thinks: "By 2040, the Sahara will be moving into Europe, and Berlin will be as hot as Baghdad. Atlanta will end up a kudzu jungle. Phoenix will become uninhabitable, as will parts of Beijing (desert), Miami (rising seas) and London (floods).

Food shortages will drive millions of people north, raising political tensions. "The Chinese have nowhere to go but up into Siberia." He hopes that "it doesn't degenerate into Dark Ages, with warlords running things, which is a real danger."

Yet Lovelock holds out a glimmer of light: "We are about to take an evolutionary step and my hope is that the species will emerge stronger. It would be hubris to think humans as they now are God's chosen race." Lovelock adds: "The human species has been on the planet for a million years now. We've gone through seven major climatic changes that are equivalent to this. The ice-ages were shifts in climate comparable with this one that's coming. And we've survived. That series of glaciations and interglacials put the pressures on us to select the kind of human that could adapt. And we're the progeny of them. And we're just up against a new and different stress. Maybe we'll come out better."

My friend Scotch-taped to the bottom another scrap of newspaper:

"The Republicans are back in control of the House, and they're bringing something with them: styrofoam cups. The cups, along with plastic forks and a number of other things seen as not eco-friendly, were done away with four years ago by Nancy Pelosi to reduce Congress's carbon footprint."
Green activists called the switch an insult to the environment, "Neanderthal" and a slap in the face to efforts to combat global warming.

My friend wonders if Lovelock is suggesting that the Pleistocene glacial fluctuations honed a more adaptable human, better able to cope with, say, the threat of modern global warming? These congressional folk not only don't believe in global warming, they think it's an environmental conspiracy. How does evolutionary pressure from the first (the Ice Age) select for the kind of person who seems indifferent to the second (climate change)? According to Lovelock, it should be the other way round. My friend finds grim humor here. Modern humans' social tendencies paint the battles black and white in a world of friends and enemies; we focus on the fights that matter the least while ignoring what matters most.

A couple more clips, intrusive but closer to the heart of the matter:

The Los Angeles Times reports "Greenland's Ice Sheet is Slip-Sliding Away." By 2005, Greenland was losing more ice than anyone expected;

the amount of freshwater ice dumped into the Atlantic had almost tripled in a decade. Summer meltwater, responding to recent warmer temperatures, also accelerated. The warm water on the top of the ice sheet made its way through a maze of tunnels, natural pipes and cracks in the ice to the bedrock below, lubricating the slip of ice over Greenland's rock basement. The meltwater descended thousand of feet in weeks not decades. This was a surprise to scientists. If all glaciers draining the ice sheet slide too quickly, they could collapse suddenly and release the entire ice sheet into the ocean.

"Should all of the ice sheet ever thaw, the meltwater could raise sea level 21 feet and swamp the world's coastal cities, home to a billion people. It would cause higher tides, generate more powerful storm surges and, by altering ocean current, drastically disrupt the global climate."

Reuters: Arctic ice sheet may swamp U.S. coasts. The loss of the huge West Arctic Ice Sheet (WAIS) would cause sea levels to rise by 21 feet in North America and 16.5 feet worldwide.

Some scientists warn that the WAIS is fundamentally far less stable than the Greenland because most of it is grounded far below sea level. One expert considers the WAIS collapse is all but inevitable given the current business-as-usual projected warming of 5-7 degrees C.

Arctica's Ross Ice Sheet is considered even more unstable than the WAIS because it had previously collapsed and could again at any moment. The Ross Ice Sheet collapse would result in an additional 15 feet of sea rise.

Fifty feet of sea rise would demolish the world's coastal populations, flooding highly populated areas such as Washington, D.C., New York City and the California coastline, and deal disaster to the low-lying Third World, displacing countless millions. And some of these rises could be extremely rapid; the collapse of the Ross Ice Sheet could cause the world's oceans to rise 15 feet in a week. If all the Arctic and Greenland ice sheets melt, the oceans would rise about 180 feet.

What about the Late Pleistocene? The most dramatic comparison to today's situation might have been the rising ocean. At the onset of the previous global warming period, known as the Bølling-Allerød, 14,700 years ago and before, when the glaciers were still at their maximum extent, the sea level off British Columbia was 300 to 450 feet lower than today. By 9,000 years ago, the ocean had risen to current levels and, after some local sloshing up and down, settled to where it is now is by 5,500 years ago.

A few scientists are now suggesting that the onset of the warming period following the last glacial maximum might have begun a couple thousand years earlier. The research, some of it from lakes in Alaska, is recent and ongoing. Future research may push that 14,700-date deeper into the Late Pleistocene, say maybe 17,000 years ago, along with earlier dates for the feasibility of navigating the Pacific Coast or the opening of the ice-free corridor. But for the purposes of this book, 14,700 years ago marks the beginning of the warming period.

That amount of ocean rise seems enormous, though moving a shell-fish camp might have been easier than relocating skyscrapers. The Pleistocene people would have seen the effects of sea rise as waters slowly drowned the forest of the Pacific continental shelf and glacial flour colored the milky deltas. They would not notice the actual rise of inches per century unless a great chunk of ice, like the Ross, broke off a sheet and fell into the ocean. But they would likely have a collective notion that the climate was changing, much as we have today.

∽

The end of the Late Pleistocene came suddenly. The global warming that began 14,700 years ago ceased abruptly with a sudden and relatively short-lived cold reversal known as the Younger Dryas (named for an ivory-colored alpine flower with a yellow center that thrived in the cooler air). This was a prolonged cold snap not a re-advance of the glaciers. Isotopes in the Greenland ice indicate the Younger Dryas (YD) began 12,880 years ago and lasted for around 1,300 years, when the warmer modern epoch called the Holocene began.

Causes of the Younger Dryas are hotly debated. One argument posits that the giant freshwater lake perched upon the surface of the North American ice sheet (Lake Agassiz) burst its ice-dam and dumped vast amounts of cold water through the St. Lawrence Valley into the North Atlantic Ocean interrupting the conveyor belt of warm surface water from the south—causing the sudden cooling. Others contend the amount of melt water from Lake Agassiz was insufficient to disrupt the heat conveyor, to the extent that it would alter climate, and that evidence of the eastward flood—flood debris, terraces or an outlet channel—is lacking. Besides, ice cores from the southern hemisphere show that the cooling was worldwide. Likewise, an asteroid theory—an extraterrestrial body smashing into the glacial ice north of the Great Lakes (but leaving no impact crater)—has played to a limited audience, and has been largely, but not entirely, debunked as the precise trigger that brought on the Younger Dryas and drove into extinction the last sabertooths, mammoths, mastodons, dire wolves, horses and short-faced bears (to name a few) in a heartbeat of geologic time.

Whatever the cause, the Younger Dryas cooling was a very big deal in America, and elsewhere; it appeared to precipitate the disappearance of the Clovis culture and their exquisite elephant-hunting spear points, along with the final extinction of the megafauna. The role of climate and human hunting on this great extinction is discussed in Chapter 9.

Worldwide, very close on the heels of the Younger Dryas, the first efforts at agriculture were germinating on an east-west axis emanating from the Fertile Crescent. Somewhere among a dozen or so places in the Middle East or Asia, someone noticed a plant she wanted to eat growing from a place where she had previously spilled wild seeds.

Another revolution was on its way, probably our biggest—the transition from hunting and foraging to farming—one in which we are still floundering, that was born of that last great blast of climate change, its progress unchecked until the burning heat of present day global warming threatens to bake agriculture out of Africa, out of Asia and banish those crops to the gulags of industrial farming in Siberia.

~

A ubiquitous stumbling block in telling the story of the Great Adventure is speculating how and if humans could have lived in North America during periods from which there is no archaeological record. Eastern and western Beringia (Alaska and Siberia) before the last glacial maximum (LGM), about 20,000 years ago, is such a place and time.

Ecologists attempt to reconstruct Pleistocene environments by analyzing ancient pollen. Beringia about 30,000 years ago was relatively mild, as indicated from lake sediment samples from Siberia, consisting of bogs and larch-birch forests amid a mosaic of tundra. About 3,000 years later, it apparently turned cold and dry.

Some caution is advised here. Scientists sampling selected lakes for spores and pollen amid a mosaic of varied landscapes don't always get the big picture right. The plant people may conclude the Late Pleistocene habitat was incapable of supporting people or animals, while at the same time paleontologists are finding fossils of big animals all over the place—suggesting the unproductive tundra looked much like an American Serengeti with its vast herds of hoofed critters. Professionals call the contradiction a paradox. The same kind of critical eye should also focus on the interpretation of the Ice Free Corridor as barren and uninhabitable (Chapter 8) or the use of fungal spores to explain Pleistocene extinction (Chapter 9). Some of the claims are specious.

Thus, scientists debate whether the tundra-steppe of eastern Beringia was too cold and dry for people or animals to survive. Palynologists studying ancient pollen cores concluded the Beringian steppe was sparse, tundra-like vegetation, more polar desert than rich grassland. But fossils dredged up by Alaskans sluicing for gold indicate an abundant animal community and contradict this notion: Bison, antelope, musk ox, mammoth, horses, bears and huge cats thrived in this landscape. Finally, botanists suggested, it might have been cold dry tundra but, unlike the mossy tundra of today, rich in grasses, sedges and forbs.

Topographically, if you subtract the glaciers, the Late Pleistocene landforms looked much like they do today. The ice sculpts the mountains into great cirques and knife-edge arêtes and, as it retreats, deposits terminal and lateral moraine that rivers outwash as broad alluvial fans. The rivers melting out of the glaciers were bigger, wider and more braided than today.

Along the southern limits of the ice sheets you might expect to find a thin ribbon of tundra next to the glaciers and along the tops of mountain ranges. Next to the tundra would be a belt of trees—spruce, fir and pine—and then temperate forests of oak, beech and hickory. Up north, the open tundra and steppes would yield to boreal forests or maybe birch and *Populus* species.

But this is only a most generalized view; in some places oak forests grew almost at the foot of the ice. Winds blew off the glaciers, picking up sand and silt from the outwash, depositing loess throughout the Midwest U.S. and elsewhere. At the edge of the big glaciers, it was windy and cold but no more inhospitable than today's Arctic.

Like our present day, species of plants and animals tend to migrate up higher on the mountain and northward as the climate warms. That is, if they can: Five-needle stone pines (like whitebark) clinging to the very tops of mountain ranges today have no place higher to go. Neither does the grizzly bear when the corridor to the next productive habitat is a valley blocked by human development and intolerance.

Maintaining corridors, wild areas and wildlife linkages is absolutely critical if we wish to save species of large animals and mitigate a few of the disastrous effects of the Sixth Great Extinction event—the one we are experiencing today.

∾

The paleozoologist Valerius Geist called our home: "The predator hellhole which was Pleistocene North America." He doubted people could have survived (they would have been eaten) in North America until about 13,000 to 15,000 years ago—or whenever the man-eaters went extinct.

Furthermore, Geist notes that not only were those predators huge, but the fossil record reveals many more specimens with significantly healed broken bones and damaged teeth than seen in modern carnivores or African species. He believes this means that the predators really had to fight hard to bring down their huge and formidable prey, that the cats and bears were perpetually hungry and desperate enough to take chances. Also, Geist argues, that the North American prey animals' "organs of food acquisition and processing remained exceedingly primitive," so that they remained in low densities and fed on only the best grass. All of which, he believes, indicates the predators were very aggressive.

One might take exception to a generalization or two, but the unmistakable point is that some experts think that Pleistocene predators precluded human colonization of the Americas until just about Clovis times. Geist thinks that the North American mega-fauna, both carnivores and herbivores, impeded human movements in two ways. The grazers created fire-resistant plant mosaics, reducing fuel buildup so that lightning produced only small fires. Humans couldn't just torch the landscape, like they probably did in Australia, and the great carnivores, the argument goes, used the pilgrims as food. Much blame is heaped at the huge paws of *Arctodus simus,* the short-faced bear.

Early Americans would have had to live with several gigantic predators, among them sabertooth cats, lions, wolves, huge cheetahs—no doubt the fastest predator on earth—and the short-faced bear. Could people have survived at all and, if they did, would those pockets of early humans have been hunted into extinction by predators, leaving little or no material record of their passing? Or of their genes?

How was it possible to live in the same valleys with this American megafauna? In addition to the short-faced bear, a number of other Pleistocene predators could have been a daily menace to these ice-age hunters. Anyone living in Beringia would have run into lions of the *Panthera* genus (the African variety but twice as big) who probably hunted in social groups. Wolves, bears and wolverines looking for an easy meal would closely follow the feline hunters.

The abundance of gigantic Pleistocene predators means a lot of killing was going on. There must have been intense competition and interaction around the carcasses of big herbivores. Short-faced bears would have challenged lions and sabertooth cats, with dire and Beringian wolves close behind, shadowed by flocks of ravens, magpies, mobs of buzzards and condors. Grizzlies were around too; probably the entire time humans may have lived in Beringia despite a gap in their fossil record from about 35,000 to 21,000 years ago, which could be attributed to a sampling bias (see Chapter 5).

With humans in Pleistocene America, what was the pecking order? Even if people managed to kill a mammoth or sloth, those other scavenging animals would be close on the scene, especially short-faced bears.

And other bears might be in the chase, though not as aggressively as the short-faced variety. Brown bears, over millennia, had learned to defer to humans, even before European firearms arrived, as told in the ethnologies of Western tribes. Early people hunted in groups, growling or roaring when advantageous; grizzlies have never been known to attack a group of four or more people (a technical exception was recorded in July 2011; seven students walking along an Alaskan river trail panicked and ran when they saw a grizzly).

Archaeologists seldom speculate about how people might have fared in such toothy neighborhoods; reasons include not only a general lack of direct evidence but also a pervasive modern ignorance about living with wild animals. North America was not like Africa where early humans and big cats evolved together—no surprises—over a span of two million years. Our hominoid ability to stand up to large predators on the savanna, even before our brain size doubled, is what paved the way for human dispersal out of Africa into the treeless north where many of the final evolutionary brush strokes to the modern mind were applied, preparing *Homo sapiens* for entry into the New World. With a large hole in the archaeological record, Beringian experts sometimes rely upon academic models of foraging for rates of human colonization.

But the first Americans encountered huge flesh-eating beasts they had never seen before, or had never seen them—unique conditions in human foraging. What might be the possibilities of sharing the landscape with *Arctodus simus*? The evidence is indirect. Did the gigantic short-faced bear, a long-legged animal that stood almost seven feet at the shoulder, limit human occupation of North America? Paleontologists and anthropological models hardly ever mention people/short-faced bear relations, though it is entirely possible that human demographics in North America could have been severely restricted by the presence of these huge carnivores.

The fact is that ice-age pioneers somehow did co-exist with some of these animals in Siberia between 13,000 and 30,000 years ago; credible pre-Clovis dates from the New World suggest that a few travelers in the lower states of the U.S. did as well. The real mystery is why don't we find evidence of many more people soaking up the sun south of the North American glacial sheets prior to the Clovis invasion?

To state it clearly, both sides of Beringia—the Siberian and Alaskan, the Old World and the New—may have presented quite different comfort zones for human colonizers. The presence of aggressive American predators in eastern Beringia, especially the short-faced bear, may refute Arctic foraging models for the whole of Beringia.

An illustration of that dissimilarity may be seen in the behavior of the brown bear.

The Eurasian brown bear and the American grizzly may look alike but their aggression levels are sufficiently dissimilar to earn the grizzly the subspecies name, *Ursus arctos horribilis.* When the brown bear crossed over the Bering Strait some 70,000 years ago to the American side, the theory goes, mothers had to protect their cubs from American lions, short-faced bears, wolves and other Alaskan predators on the open tundra. The best defense was a good offense. Grizzlies charged and, when necessary, attacked threats to their young.

It might be informative to examine the possibility that Pleistocene North America might have been an unusually rough place to live. The presence of all those predators could have squeezed human consciousness into a tight focus that could shed light on the astounding and explosive nature of American colonization around 13,000 years ago.

$\sim$

A seminal moment in the life of a hunter arrives when he finds himself the hunted: That dread second when he is frozen in his tracks at the edge of the meadow by the eerie silence in the forest; he feels a primordial but familiar tenseness clamping the back of his neck and he realizes that he is being stalked as prey by a large beast.

This ancient relationship doesn't present itself to the modern world as frequently as it did prior to the industrial age or, especially, during the last days of the Pleistocene. In fact, predation on human beings is so uncommon today that when a single lion, bear or tiger emerges from the bush to stalk, kill and sometimes eat a human it generates international news and best-selling books. A much-chronicled modern account tells the story of a Siberian tiger's vengeful attack on a man named Markov, a poacher who had previously hunted and wounded the huge cat. The vendetta took place during December of 1997 near the Amba River in the Bikin River drainage of Russia's Far East. This predatory tiger incident was first chronicled in 1998 by renowned Russian bear and tiger biologist Dmitri Pikunov; the details of this particular attack, however, constituted such a good story that they were rewritten into a popular book in 2010.

The author of *Tiger* provides a few details of the attack: "In 1997, the Russian investigating officer who was tracking the tiger, reported: '(he) had never seen a fellow human so thoroughly and gruesomely annihilated. It looks at first like a heap of laundry until one sees the boots, luminous stubs of broken bone protruding from the tops, the tattered shirt with an arm still fitted to one of the sleeves. Here, amid the twigs and leaf litter in the deep Russian forest, not far from his small cabin, is all that remains of Vladimir Ilyich Markov.' The tiger who killed, dismembered and ate Markov waited for him a long time, perhaps days, lurking near the door of his cabin…."

This huge male tiger had previously destroyed everything that had smelled of Markov, and then waited for him to come home. The attack seemed chillingly premeditated.

About this time, as I read on, a chill ran up my own neck. Something about this tiger sounded familiar. *How old was this cat?* I reread the book but all I could definitively glean was this was a very large male tiger. I think male Siberian tigers, like male grizzlies, continue to grow in size with age. Tigers can live to be 15 years old or so in the wild, although large tigers tend to be targeted by poachers and are therefore rare. The tiger who ate Markov was later killed but never weighed. An experienced eyewitness said he had "never seen a tiger as big as that one."

A male tiger maintains an exclusive range, driving younger males away or killing them. Siberian tigers have huge territories. Could the killer tiger be ten years old? Possibly. I do the math. Probably, I think. Dmitri Pikunov would know for sure. The book says Dmitri has had a serious heart attack or I would ask him directly: Is the killer tiger the same one we trailed in 1992? We crossed the tracks of the tiger in question four miles southeast of the Markov attack site.

∼

Dmitri Pikunov and I were 2-person tent mates on a kayak trip down the wild upper Bikin River in 1992. At least I think it was 1992. I dig out my field notebooks: Yes. Our journey was a buddy trip with five American friends: Jib Ellison, Doug Tompkins, Rick Ridgeway, Tom Brokaw and Yvon Chouinard—famed kayakers and mountain climbers, well, all except myself and perhaps Brokaw. We spent about three weeks in Siberian tiger country, the last ten days fishing and paddling down the wild Bikin River.

We ran into Dmitri in Ternai while struggling to break loose of the Russian bureaucracy and get into the wilderness:

In order to visit the countryside, we are told, it is necessary to secure a permit from the Bureau of Tourism. The Director of Tourism offers us a river trip using our own kayaks for only $2,100, American.

"A truck and motor boat will accompany you at all times," he says.

This is not exactly what we had in mind. I stare out the window of what until recently used to be the Communist Party building: A pretty girl is walking her cow down the street.

"This is banditry," says Brokaw who along with Jib has acquired the unsolicited job of group-diplomat. We are getting nowhere. Jib stands up and announces that "We are out of here, we are going home."

These guys are good sports, they roll with the punches and there is no whining.

By fortune, we run into Pikunov. He knows we are interested in preserving wild country. Dmitri's greatest personal accomplishment, he tells us, was in helping establish a Native People's Reserve in the Bikin for the Udege people. The Bikin River country, he argues, is "the most beautiful, most pristine of all."

"You must see it," he says. "Hyundai wants to cut it all down and Moscow will cave in to them."

The die is now cast. We decide to ignore warnings that we must get permission from the KGB to travel: We will try to bribe a helicopter pilot on our own to fly us and our fold-up kayaks into the headwaters of the Bikin River. It can be done, we hear.

We dig into our pockets and come up with a roll of cash that we pass to Dmitri. We find a chopper. Dmitri Pikunov says, "Speak no English." He covertly passes the roll to the pilot. Soon we are airborne.

We have a single map. The country is huge with no trace of man upon the land. The map shows the middle tributary of the Bikin River, the Zeva River, unfurling counterclockwise, flowing through eroded volcanic hills and cliffs of columnar basalt, finally hooking into the Bikin. That's where we want to go.

Yvon and I look out the open window of the big military-style Aeroflot helicopter, the port that Rick has opened in order to take some photographs. As our only map of the area is passed back to me, I stupidly grab it in front of the window. In a heartbeat, half the map—the half that shows the Zeva and all the country we plan on kayaking—rips off and is sucked out the window. We are now map-less and I wonder what my carelessness portends.

The boys, especially Tom, will later make me pay heavily for this blunder as the two of us vie for "worst" in kayaking skill.

Nonetheless, rivers tend to run downstream. My journal tells of the last days of our trip, when we walked up the Amba River:

> "Amba" means tiger as well as "devil" in the native Udege tongue. We beach at the mouth of the Amba River and walk upstream a short way to a trapper's cabin (which belonged to the "key witness" in the Markov incident) that Dmitri used in past years during his study of bears and tigers here. The Amba River bottom in summer is hot and humid. Dmitri leads us on a hike several miles up river. Shoulder-high cone fern and alder obstruct our vision. Moss and shelf fungus grow on logs and downfalls. During the winters of his bear study here during the late 1970s, Dmitri would ski along the river and bang on cottonwood trunks with a heavy mace, waking the Asian black bears that hibernated within the hollow trees. The Amba is also prime tiger habitat. China lemon vine grows on the smaller trees, and cow parsnip and nettles make up the under-story. Ticks hang off the low vegetation; we stop for a tick-check every fifteen minutes. This is our last day in

the wilds; tonight, we paddle on down to the big Udege village where we can hire a truck to haul us and our fold-up kayaks out to the Trans-Siberian railroad.

We climb a steep hill to a ridge. There is wild boar and black bear sign everywhere. Dmitri signals for us to be quiet. Our crew is noisy, distracted, self-absorbed, talking of industrial collapse and geopolitics. Dmitri snaps at us to shut up. We can hear movement down the ridge. Up ahead we hear the breathing sounds of big animals—probably bears huffing away or boars snorting, all now running downhill.

We blew it. The world is only as big as we allow it to be. Wild places and animals pass along their secrets only if we listen. You have to pay attention. A touch of danger would help. You need to know you can die: A surprise rapids the size of Lava Falls, a bad stretch of black ice across an ice chute, a white-out on a glacier, or maybe a bear or, especially, a tiger. But it's hard here on our last day out before the slow return home. It's especially hard in a group; the social dynamics can drain you of vital curiosity and attentiveness.

I split off by myself for a short time. Asian black bear have ripped branches off trees everywhere. I find day beds of boar and bear; there is sign of digging around the large Korean pine trees. The big live oaks are lovely. It's good to be off alone; I find a bear-ripped honey tree and an ancient yurt on top of the ridge—built by either an Udege trapper or Chinese ginseng hunter.

Dmitri signals for me to rejoin the group. We drop back down to the Amba bottomland, finding an old trail.

Suddenly, Dmitri freezes and motions me forward: A tiger track glistens in the mud. The track in the wallow appears to be only about a day old, around five inches across—the print, Dmitri says, of a young but dominate (about five years old) male cat that has replaced the previous dominate male cat, who was killed by a poacher. The young tiger leaves scrape marks every few hundred meters and spray scents on territorial tree markers. We stop at such a tree. The bark has been rubbed off by Asian black bears who also are attracted to the strong scent. I get down on my knees and press my nose against the bare trunk.

The pungent fetor of tiger fills my nostrils and—for just a second—I travel with the big cat, orange and black stripes flashing barely perceptibly through the sea of green undulating cone fern, into the wild and predatory world that not so very long ago was my own.

If the huge male tiger who killed, dismembered and ate Markov in 1997 was ten years old, quite likely it was the same then-younger cat whose scent we snorted in 1992 on the Amba River. We were, at that time, less than five miles away from the Markov attack site.

Somehow, this apparent coincidence didn't hit me as startling: The fact that we probably crossed the sign of a tiger who later tore a man to pieces and ate him, for me, many curious precedents. Deep in the Sierra Madres of Chihuahua, Mexico in 1985, a jaguar coughed and sprayed just beyond the light of my campfire. It was my first jaguar and I glibly imagined this rosette-spotted predator stalking me. The next morning I discovered the jaguar had backtracked me for 14 miles.

And I remember vitality at the edge of fear infused into my own life when my mountain campfire was besieged by a huge black grizzly—he knew me—one stormy autumn night near Glacier Park in Montana: Only a few days earlier the same bear had ripped a cache of camping gear from a tree and had chewed to bits my sleeping bag and sweaty T-shirt, everything that smelled of me, while ignoring a tent and other gear which did not (sending me an unmistakable message: "Get the hell off my mountain." I did).

The sentience of large predators is unlike what you see gazing into the orbs of a chimp or your favorite Labrador retriever. The tiger with a vendetta or bear with a memory stirs a different set of sensory responses cached deeper down in our brainstems. I knew a grizzly in Yellowstone, one I tracked for a decade, who set up what looked like a deliberate ambush for me in the snowy woods. I had snowshoed into a remote thermal area and later found a huge male grizzly on a winterkilled bull bison. About dusk, the bear rose and followed my snowshoe tracks out onto the crusted snow into the timber. I waited ten minutes and started after him. A few feet into the darkening forest, I stopped and looked at the huge, twisted paw imprinted over my own snowshoe track. A premonition rose up my spine to the hair on my neck. I retreated rapidly and pitched a distant tent in the darkness. The next morning, I cautiously followed the skewed tracks: the big grizzly had trailed my snowshoe prints for a hundred yards, then his tracks veered off sharply in a tight circle that led to an icy depression in the snow behind a large deadfall ten feet off my trail. Had I gone any further in the darkness, he would have been right there. The icy bed spoke of a long wait.

Years later, I investigated a bear mauling: In August of 1984, a woman camper was killed and partially consumed in the backcountry of Yellowstone's upper Pelican Valley. In late October of that year, I hiked back to the site of the fatality. I squatted to fill my canteen at a small pond. I tensed feeling a sudden tightness in my lungs, then a crushing pain in my chest: In the frozen mud I saw the distinctive track of the big grizzly who would have ambushed me eight years earlier. For a moment, I was disorientated. The tracks were old, unconnected to whatever happened here in August. He probably wasn't the killer. But he had been right here. I hadn't known what to make of the snowshoe ambush in 1976 either. Maybe the grizzly was just curious. I felt separated from the magic that once connected me to the grizzly with the crooked track. The authorities never found the bear who killed the young woman. I left the site of the fatal mauling bewildered, isolated from my own kind, wondering how ancient people maintained their humanity among the other nations of animals. An acting park superintendent put out an odd statement: "The last thing we want out there," he said, "is the legend of a killer grizzly."

The point of these recollections is that sharing the wilderness with legendary killer bears or cats dramatically shifts the psychic landscape. You think about the world differently because you have no illusions about being in control. If you traveled armed with a spear in Pleistocene North America, you definitely lived in the middle of the food pyramid, stumbling about like a minor but tempting pork chop, hunted by the big cats and bears while you pursued the mammoths and camels of your ice-age vision quests.

∽

The most mercurial player remains the gigantic short-faced bear who was a matchless American native. What kind of creature was *Arctodus simus*: Was it indeed the terror of the tundra, as suggested by a few eminent paleontologists, or a peaceful grazer of the plains?

Considerable academic debate rages about whether the giant short-faced bear was a practicing predator, scavenger or vegetarian. The more interesting question is what sort of challenge *Arctodus simus* presented to people by appropriating the hunters' kills and in actually bringing down people as prey. Paleontologists are not a lot of help on this one; they tend not to directly address such human/bear issues.

The core is this: How does it feel to share the land with creatures who are aggressively trying to kill and eat you? It probably felt like it does today, regardless of the rarity of modern human/predator relationships. The hunter who roams the land, spear in hand—looking over his shoulder for the bear or cat that, indifferent to our emerging dominion, regards puny two-legged *Homo sapiens* as just another variety of pot roast—carries with him a valuable awareness of vulnerability that we lack today in our safe, sterilized woodlots of well-managed whitetail deer and high-powered rifles. That value, I believe, lies in perceiving authentic risk that in turns triggers an appropriate survival response. In such a cosmos, you hear more, see more and smell more. Today's polluted global winds dilute the olfactory discernment of the shadow of the sabertooth. I'm saying maybe we could use some of this biological insecurity and acuity in today's world.

It is within that context I consider the short-faced bear.

∾

I would think the short-faced bear potential to limit early American demographics too important an issue to be ignored by either archaeologists or paleontologists, although, for the most part, the academics avoid specific speculation. I searched the professional literature for clues and illumination. The venerated vertebrate paleontologist Bjorn Kurten once called *Arctodus simus* "the most powerful predator in the Pleistocene fauna of North America." Here is a vulnerable windmill, like "Clovis First," that makes for great press and it's something to tilt against. Paleontology often does a good job of debunking popular misconceptions. Archaeology scarcely mentions human/short-faced bear interactions. Paleontology does so only indirectly in a few technical studies—

the sort that I endeavor to steer clear of in this book. The reasons are simply to avoid getting snagged in technical detail as well as an acknowledgement of my layman limitations.

Nonetheless, I reluctantly grabbed a handful of recent papers as an example. The representation is probably biased because the ones that caught my attention were either very enlightening or exasperating. They are, however, about the only source for beginning a discussion. The scientist's conclusions are examined in a stew of my own experiences with grizzly bears. Here's a summary of three nearly random, technical articles that help paint the panorama. All challenge *A. simus*'s reputation as a predator.

Spanish paleontologists measured a small number of bone pieces and fragments of short-faced bear fossils and inferred that the bear was not short-faced, long-legged or predacious. One question, which was never asked, is what determines what a bear eats? The answer is behavior, especially aggression, not snout size nor the cut of their omnivore teeth. Animal protein is universally preferred over vegetation. Aggression and dominance played a huge role, especially around the kill sites of Pleistocene carnivores. Predation is opportunistic.

The study purported to compare *A. simus* to the grizzly bear implying the short-faced bear was a slow moving vegetarian. But grizzlies can outrun racehorses over a short distance and bring down adult elk, caribou, moose and the calves of all these creatures. The short-faced bear evolved in an America without people—grizzlies did not. From my own observations, brown bear routinely displace wolves, cougars and, less commonly, humans from carcasses, presumably because the bear has reason to fear humans. The short-faced bear had no reason to fear *H. sapiens* because it had never seen one until the late autumn of its species some 13,000 to 30,000 years ago. Since there is no record whatsoever of human interaction with any of these big prehistoric carnivores (there is an anecdotal rumor of a New Mexico Clovis point lodged in a dire wolf's jaw), my account of such relationships is speculative, based on my own experience with existing American carnivores—polar bears, wolves, cougars, jaguars and brown bears—the most dangerous of which, statistically, is the grizzly (it's a misleading stat; moose are more of a threat to people).

But all these modern beasts are pussycats compared with those extinct predators.

Other paleontologists studied the teeth, along with some skeletal morphology, of *Arctodus simus* and concluded the bear's diet largely consisted of coarse foliage by unselective grazing. I ended up wondering how any bear could survive the Beringian winter by unselected grazing of coarse foliage.

Sometimes paleontology raises important qualifications in characterizing *A. simus* as the once-dominant predator in North America. A useful study suggests that because the short-faced bear was incapable of sharp turns or of stopping on a dime (based on examination of fossil skeletons), it was therefore not a fierce or pursuing predator. The bear's gracile bone structure argued against wrestling with a mammoth or giant sloth. Having had the opportunity to handle a museum specimen myself, I couldn't agree more that this big-hipped, broad-nosed bear with huge crushing jaws was a superbly equipped scavenger. On a continent crawling with giant predators and prey, carcasses would have been commonplace. The short-faced bear would certainly be a main contender for any kill by any animal within its olfactory range. The large hipbones hint at an animal that could, like a grizzly, stand and scent carrion from several miles away. Modern grizzly bears have been recorded scenting carrion at a distance of nine miles. Standing on its hind legs to reach fifteen feet tall, with its wide nostrils flaring, the short-faced bear might have been capable of smelling a carcass at a much greater distance. (Incidentally, ethyl mercaptan, the essence of rotting meat, is also markedly detectable by modern human beings, hinting at a scavenging past life—early hominids driving jackals off a zebra carcass and then pigging out on the rotting flesh—no doubt a key to our early survival on the savanna. This is something to keep in mind when we recast early Americans, or Africans, as robust big-game hunters: We were also adept scavengers.) These scientists also implied that its great size and ability to stand made the short-faced bear an intimidating presence (no doubt, but it was also well equipped to fight off other big carnivores during the violent strife that inevitably occurs at kill sites) near carcasses, further demonstrating that the big bear was an accomplished scavenger but not a formidable predator. This may be true.

But, collectively, these sorts of studies are misleading. They tend to be tight-lipped about human/short-faced bear relationships; this discussion overlooks the fact that the giant bear might have been a considerable threat to other mammals, especially two-legged ones. Here is a pitfall in tracing the path of the first Americans. Paleontology (in this case), like archaeology, is an interpretive science. This is what makes it so interesting; it's not experimental or a straight recitation of facts. Taking on such vulnerable precepts as "Clovis First" or the gigantic short-faced bear as "the most powerful predator in the Pleistocene fauna of North America," makes for great press. But large scavengers, like hyenas, humans and brown bears, can at times also be formidable predators: The likelihood of one pattern of behavior does not preclude the possibility of the other.

Short-faced bears belong to the *tremarctine*, or "running bears," group of New World bears. Having spent much of a lifetime living close to wild grizzly bears, I don't buy into a demystified edition of *Arctodus simus* as a peaceful grazing omnivore. With big bruins, behavior is what defines what a bear wants to eat and how he gets it.

Short-faced bear/brown bear relationships are discussed in Chapter 5. Grizzlies are true omnivores and prodigious diggers with long claws. Adults can't climb trees. Brown bear hibernate; they dig dens on high mountain slopes.

With its great body mass, the short-faced bear couldn't make it up a tree and likely didn't dig much either. Many paleontologists doubt *A. simus* hibernated. If they could find sufficient protein during the winter by scavenging or predation, they didn't have to. Smaller specimens of the short-faced bear have been found in caves in the contiguous states. These caves are probably not normal lairs but rather, in a species with pronounced sexual dimorphism (male bears are larger than females), the winter homes of pregnant females or mothers with young.

In solving the argument about what the big bears ate, a key observation is that the grizzly survived the Late Pleistocene and the gigantic short-faced bear did not. The obvious reason, I think, is that the brown bear is a true omnivore, more flexible and adaptable than the meat-eating *A. simus*. Once the great megafauna declined, the serving portions for scavengers and predators certainly shrunk. If the short-faced bear had been a successful non-selective grazer and browser of shrubs for fruit, it might still be around. Bears can't live by non-selective grazing. If the short-faced bear's broad muzzle, as argued by some of the paleontologists, precluded selecting green grass and sedges from more cellulose-laden forage, then *A. simus* could not have plucked berries from the bushes they grew on. That makes for a lousy vegetarian. Also, *A. simus* didn't den, as far as we know, and would have had trouble finding den sites with caves rare and their claws poorly designed to dig dens. The big bears died off because they couldn't live off plant life when their real food, the great megafauna, dwindled down toward extinction 13,000 years ago.

In the far north, the short-faced bear diet, confirmed by isotope studies (the nitrogen-15 signature is that of a carnivore) on fossils, is meat by predation or scavenging. Any sizeable carcass in bear country means competition by violent defense and fierce aggression. The notion that the short neck of *Arctodus simus* evolved as a consequence of the need to support its massive head seems misplaced. What comes first? Nature doesn't evolve such long limbs without selection for some function of speed (which might also have made the bear appear more intimidating at carrion sites). The swiftness of this predatory beast—and this is the real point—was sufficient to bring down lumbering (compared to antelope and jackrabbits) humans and other slothful creatures in open country.

The kinds and amounts of vegetation any bear species can digest depend on the elongation of its basic carnivore gut and there is nothing unselective about a bear's grazing. In fact, living bears are among the most selective of omnivores in choosing only green vegetation in its pre-flowering stages. In winter, when only indigestible cellulose is available as plant food, the bear must eat meat or hibernate. The ice-loving, carnivorous polar bear is, of course, the exception.

Never having seen an upright primate before, the short-faced bear could have been a very serious problem for two-legged immigrants especially in the American West. Unlike grizzlies it had no prior reason to fear humans. The now extinct bear was certainly capable of snagging and eating a wandering clansperson. *Arctodus simus* might have appropriated the kills of early American hunters depriving them of their food. People living south of the ice before the Clovis arrived were in part big game hunters as evidenced by documented mammoth kill (or scavenging) sites. Clovis hunters certainly were. Mammoth, like most of the huge herbivores, were open country animals; how did people secure their grassland kills against this giant scavenging bear? South of Beringia, paleontologists think the short-faced bear preferred higher, well-drained grasslands mainly west of the Mississippi River. Until recently, no giant short-faced bear fossils had been found in the southeastern U.S., but we now have specimens from central Florida and Virginia. It might have been easier for people to survive in the eastern woodlands of North America where, perhaps not coincidentally, a few of the more credible pre-Clovis radiocarbon dates derive.

∾

One can make an informed guess at the hunting opportunities for humans and animals. Except for glacial landscapes, the topography in which humans and extinct animals lived remains essentially the same as today and we know something about ancient vegetative communities because they still grow to the north or at higher elevations, in slightly altered plant communities. To evaluate the human/short-faced bear relationships, however, we might reduce those habitats to just two: open country and forests. Open country would include tundra, steppe communities, the plains, high deserts, chaparrals, savannas and the American Southwest.

Pre-LGM or pre-Clovis people living in open country would have had a challenging time guarding their camps and securing their kills against marauding scavengers. Unless these giant bears were sated from a super-abundance of Late Pleistocene predator-killed carcasses, they would have likely come sniffing about camps and butchering sites. Valerius Geist's observation about healed broken bones and cracked teeth in the fossil record of the megafauna speaks against this full-bellied possibility—suggesting the predators were persistently aggressive. The short-faced bear, with a nose like a wolverine (who can scent a mountain goat carcass from twenty miles away buried ten feet deep by avalanched snow) could ferret out the most carefully concealed meat-caches of early American hunters.

In the eastern forests of ice-age America, short-faced bears were less common and humans might have had a better chance at survival.

Besides fire and social cohesion, what weapons were available to early Americans for defending their kills and camps from short-faced bears? Here we have a preview of the chapters that follow: There is no archaeology of the earliest people who might have visited America before the last glacial advance. We suppose they had bolos, slings and spears of wood, perhaps tipped with bone points. Just before the Clovis technology shows up in the contingent United States, a smattering of sites reveal mostly small flakes and points but nothing with which you'd want to face *A. simus*. Clovis technology is different; Clovis projectile points are big and sturdy bifaces used to spear mammoth. These heavy six-inch spearheads might have allowed hunters to drive the short-faced bears off their kill sites whereas earlier people with smaller-tipped weapons remained at the whims and mercy of the big bear. Also, the *Arctodus simus* population could have been on the decline from natural causes related to climate by Clovis times.

Another possibility is that later pioneers used dogs to keep short-faced bears at bay either at camp or while butchering a kill. Recent genetic studies suggest that American and Old World dogs descend from a single domestication event in China around 15,000 years ago. But wolf-dog skeletons show up at older archaeological sites (one from the Altai Mountains of Mongolia dated older than 30,000 years) and it seems reasonable to assume there should have been multiple origins for man's best friend. No evidence has been found that proves the earliest American hunters used dogs, but it makes sense.

A final question, perhaps the most important, is exactly when, and where, did the gigantic short-faced bear population decline and succumb to extinction? The big bears might have gone out in different parts of the continent at different spans of the calendar dates. The final demise of the short-faced bear, sometime about 13,000 years ago, and the explosion of Clovis across the lower 48 states seem nearly synchronous. But a few of the big bears may have lingered. From Kansas, conflicting dates of 9,630 and 10,921 radiocarbon years are listed for the same specimen. The most reliable terminal date might be 12,800 years ago from Utah. No doubt a few stragglers, for which we have no record, hung around for a while, but as a functional threat to humans, the bear was on its way out.

What's the last date for a short-faced bear in geographical association with Clovis artifacts? In 2010, I looked into locations of *A. simus* fossils for a talk I gave in New Mexico and found that there might be some overlap of Clovis and *A. simus.* A few scholars think the demise of the short-faced bear was a precondition paving the way for the Clovis invasion. I found many (at least a dozen) museum specimens of *A. simus* from the Southwest that are poorly dated or undated—simply listed as Late Pleistocene because they were recovered before radiocarbon dating came into use. Most fossils came from caves and at least a half dozen of those records list a Clovis projectile point from that same cave.

Did Clovis hunters actually co-exist with the last of the short-faced bears or did the final extinction of *Arctodus simus*, in this case in New Mexico, allow the unprecedented explosion of Clovis culture across the American landscape? Here is a crucial question that we might answer by dating and re-dating those existing specimens.

❧

Some archaeologists mention poisonous snakes or plants as impediments to colonization though these deliberations seldom go into much detail. These dangers are often lumped in a larger conversation about landscape learning, which can be a very rarefied field of models, extrapolations and exquisite graphs. Their practical use in the archaeology of early Americans, however, has to do with how fast people might have traveled or migrated. Would the presence of New World poisonous critters or plants have slowed people down? How you feel about that might be influenced by what you think about the Clovis invasion—whether they blitzkrieged across an empty continent endlessly chasing mammoths over the next pass or if they spread their signature artifact among a pre-existing American population who had taken ample time to settle into the environment, slowly learning about the plants and animals over many generations and several millennia? I'm saying it's possible that this might not be an entirely objective discussion.

So, would the poisons in snakes and plants (and other less-deadly organisms, such as toads, arachnids and insects) constitute a significant threat to early American colonizers? The principal candidates include paralytic shellfish disease (PSD), the poisonous toxins produced by microorganisms associated with "red tides," poisonous plants, mushrooms and snakes. And, even if early foragers commonly encountered such species, wouldn't these endangerments pale in comparison to dangerous beasts, avalanches, blizzards and river crossings?

Paralytic shellfish disease is associated with warmer and calmer waters than those encountered by the first Americans, who either boated or walked ashore at the Bering Strait and were therefore familiar with coastal habitats. In either case, some of their ancestors would have known of PSD, which is discussed in Chapter 6.

Deadly plants or mushrooms are almost a non-issue throughout much of North America. There are so very few truly dangerous species, such as water hemlock and whitish Amanita mushrooms ("death caps"). Modern travelers have mistaken water hemlock for cow parsnip or Indian celery (*Heracleum lanatum*) whose young umbel flowers are heavily grazed by bears. The single case of people eating water hemlock of which I have personal knowledge of, occurred at Heart Lake in Yellowstone Park, and the attending backcountry ranger was a friend of mine. It was a terrible, agonizing death, the kind that would have stamped the most cautionary message on any band of witnesses, including Late Pleistocene hunters.

Most species of "poisonous" plants and mushrooms in North America are merely inedible due to foul taste or make you sick with intestinal symptoms. You don't, however, die from eating them unless you are severely allergic or insist on eating a belly-full. Some inviting appearing species that are moderately poisonous, such as baneberry, give immediate warning by their taste—exceedingly bitter with a burning aftertaste.

Large, attractive and tasty (death-bed accounts have left us this useful information) whitish Amanita mushrooms, especially *A. phalloides* and *A. ocreata*, could have knocked off a few early explorers. The symptoms don't show up for a day or three, so the linking of the poisoning to the eating the mushroom might be missed. Before the end of a week, the Amanita toxins destroy the liver and kidneys. There are other deadly types such as *Galerina spp.*, but these are tiny and unappealing fungi. Early Americans were largely hunters (and likely scavengers); I'm not suggesting these ice-age people would show an inclination towards gathering much beyond the most accessible fruits and berries, but mushrooms might be a special case. Ethnologies of traditional people reveal a sophisticated interest in mushrooms as medicine and psychedelics. The earliest archaeological record comes from rock art sites: Prehistoric people painted bell-shaped caps of "magic" mushrooms of the genus *Psilocybe* on cave walls 9,000 years ago in the Sahara and 6,000 years ago in Spain. The famous Iceman of the Tyrolean Alps stashed dried polypore fungus in his medicine bag 5,300 years ago. Whether that indulgent-appearing curiosity extended back into the American Pleistocene is unknown. The draw to experience altered consciousness through psychedelic plants and mushrooms may be closer to an innate human inclination than the DEA ever imagined.

Having spent decades in pursuit of wild edible mushrooms, I have sampled about five-dozen species. My usual approach to a new mushroom—my limited expertise often falls far short of precise species identification—is to smell, taste, cook and then eat a tiny portion of the unfamiliar fungus. If all goes well, I wait a day or two to see how it agrees. The first rule in wild mushroom gathering is to learn the deadly ones. The whitish (sometimes yellowish) *Amanita* species are the bad ones. I learned about deadly amanitas from books. Ancient people had to learn from trial and error. The acquired knowledge that certain mushrooms were poisonous would have been an expensive one to the band, gradually learned, painful and not easily forgotten.

In addition to poisonous plants, some archaeologists make a case for rattlesnakes as animals that the earliest American travelers had previously never encountered before and would therefore have had trouble adapting to. This again is misleading; poisonous snakes have been our companions since the first hominoid stood upright. Beringians, having made it to the Arctic by way of mid-latitude Asia, would have had all kinds of experience with and folklore about pit vipers—the most common group of North American poisonous snakes, which includes rattlers. As a Green Beret medic in Vietnam, I successfully treated five snakebites from Russell's and Malaysian pit vipers, serpents that look and act just like our American rattlesnakes without the rattles. The rattles just make our American pit vipers easy to locate.

At many times in their history, ice-age people would have known and learned about these reptiles, plants and mushrooms—a collective oral tradition of caution.

∾

The great ice sheets of North America presented a formidable barrier to human migration from Beringia to the lower 48 states. Yet routes southward and corridors through the ice were open before and after the LGM.

Poisonous plants, mushrooms and reptiles did not constitute a significant barrier to human travelers in North America.

Other animals, however, could have constituted a major deterrent to early colonization of the Americas. Among the Late Pleistocene megafauna were several huge cats and big wolves. The American lion was twice as big as its modern African cousin and probably hunted in social groups. A couple sabertooth cats and a large cheetah were also on the scene. Any of these predators were capable of snagging a careless human traveler. Wolves were probably more aggressive during the ice-age than those still lingering in our northern woods today, especially at kill sites and around carcasses. It would have been dangerous to try to drive off either the dire wolves or the big Beringian wolves.

The quintessential North American carnivore was the giant short-faced bear, *Arctodus simus*. Though mostly, paleontologists think, a scavenger of kills by other predators, this huge bear could have run down vulnerable humans in open country, much like grizzlies do with elk, moose and caribou during favorable seasons. The most formidable challenge for early American migrants, however, would have been securing their kills from the depredations of the short-faced bear. In the far North, where big game hunting was a necessity for surviving the long winter, *A. simus* could have precluded significant human occupation. South of the ice, the huge bear might have made big game hunting, especially of mammoth, next to impossible: *A. simus* could pick up the blood scent from miles away and aggressively appropriate the carcass. Prior to the decline of the megafauna, including the scavenging short-faced bear, pilgrims south of the ice might have had a better shot at surviving by foraging in small bands and hunting small game in the forested parts of the continent. *Arctodus simus* was still around as late as 13,000 years ago but perhaps in declining numbers. The demise of the gigantic short-faced bear may have opened a very big door for the colonization of the Americas.

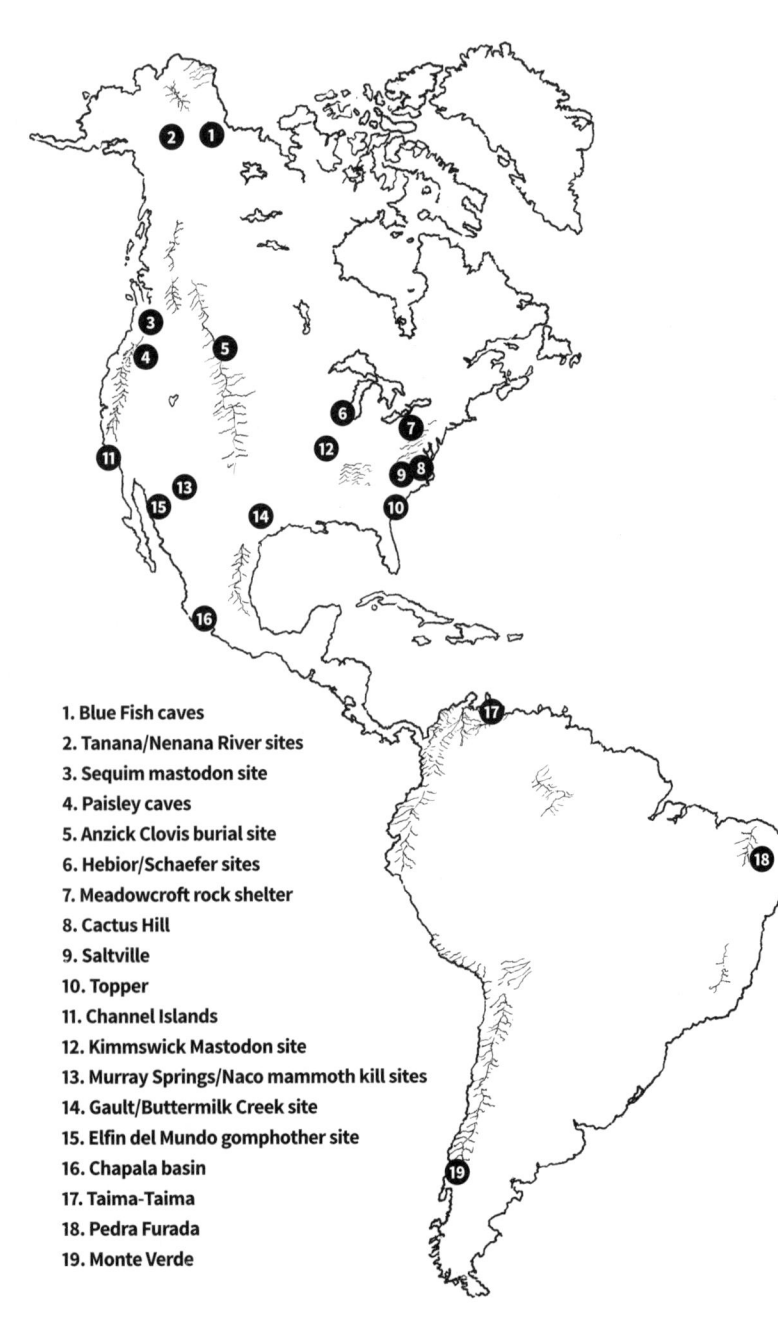

1. Blue Fish caves
2. Tanana/Nenana River sites
3. Sequim mastodon site
4. Paisley caves
5. Anzick Clovis burial site
6. Hebior/Schaefer sites
7. Meadowcroft rock shelter
8. Cactus Hill
9. Saltville
10. Topper
11. Channel Islands
12. Kimmswick Mastodon site
13. Murray Springs/Naco mammoth kill sites
14. Gault/Buttermilk Creek site
15. Elfin del Mundo gomphother site
16. Chapala basin
17. Taima-Taima
18. Pedra Furada
19. Monte Verde

CHAPTER 3

Archaeology and the Shape of the Journey

THE PARAMETERS OF THIS STORY have been delineated by a multitude of disciplines including paleontology, geology, palynology, zoology, genetics, linguistics and Pleistocene ecology. But we can especially thank the field of archaeology for providing the major theories and framing the principal arguments that rage around human migration routes into the Americas and the causes of the Late Pleistocene extinction. This book will not feature dueling archaeologists, though the professionals do an admirable job of duking it out on their own in the academic journals and their own books. This material is not difficult to read and we live in a time when most all the hard-earned research is available to the layperson. I am grateful for this scholarship as the bedrock matrix from which to explore the decidedly less-scientific areas of the human spirit and its adaptations to change. Even the most technical of this mountain of research papers is anything but dull.

In fact, the origins of the first Americans and the extinction of the Pleistocene megafauna, including the role of climate change, are among the most attractive areas of American archaeology and exceedingly popular subjects in the mainstream press. You can read about it in the *New Yorker* as well as *National Geographic* and it's all over the science and nature channels.

The story of these first American adventurers is perhaps the most prestigious scientific area of New World anthropology. Indeed, within the often-insular world of early American archaeology, no issue is more controversial, important or glamorous than detecting the origins of the first people in the New World and uncovering when and how they got down here from Asia or across from Europe. Where did Clovis come from? At the end of the Ice Age, the huge animals disappear forever and we see the abrupt end of the Clovis culture. The questions surrounding these issues represent America's greatest unsolved archaeological puzzle.

~

The shape of the archaeological journey begins in temporal mystery. Some time beginning or after about 30,000 years ago, a group of Siberians (most authorities believe the first Americans came from northeastern Asia) squinted into the Arctic sun that rose over the largest human uninhabited landmass in the entire history of exploration. For the first time, they glimpsed the snow-covered highlands of western-most North America, a continental expanse that constituted the longest empty frontier ever encountered in human colonization. No people lived in the two Americas—some 16 or 17 million square miles without a single footprint of *Homo sapiens.*

Of course the Siberians didn't know this. They were hunters and they saw the tracks of familiar animals coming and going across the lowlands later known as the Bering Strait. This was not a casual crossing: it is notable that two huge creatures known to early Beringians, the Old World woolly rhinoceros and the American gigantic short-faced bear, stayed home, while horses, camels, mammoth and humans apparently crossed with some frequency. For two-legged hunters it was probably a matter of following animal tracks. Just as you would not explore a remote thermal area of Yellowstone Park, where you can step through a thin crust of sinter into a boiling death, without looking where the bison had walked, you wouldn't venture across a wide stretch of ice or a strait the size of Bering—if they knew it was a bridge—without the insurance that a large animal had preceded you.

The Siberians might have waited out the winter darkness and made their initial move during early spring, when they had daylight and a few months of summer to change their minds in case the New World didn't work out. But at some point, these people headed east and crossed the strait on ice or on the dry land bridge that would have been available several times during the major glacial advances of the last 50,000 years. This opportunity ended with the global warming period that began to melt the continental glaciers about 15,000 years ago and subsequently flooded the route four thousand years later.

When the ice-age hunters finally crossed, what they found on the Alaskan side was enough to hold them in America. We know little of these early people, how they moved around or how they survived. They were most certainly seasoned ice-age travelers who followed their prey animals, who knew how to make fire, clothing and movable shelters. We also know these Siberians hunted big game—mammoth, bison, reindeer and horses—as well as hares and birds in frigid habitats far north of the Arctic Circle 30,000 years before the present. They could travel over ice and snow, eventually with sleds pulled by dogs, whose domestication stretches back at least 15,000 years. In short, they were amazingly well adapted to this severe, unforgiving environment.

And so they arrived in America. Most remarkably, they found no one. No people were out there in front of them. Eventually—maybe right away—the first Americans would have recognized that they were alone: No tracks on the beaches or snowfields and no campfire smokes on the horizon. Every so often, these early hunters would encounter animals and plants they had never seen before. The part of North America these Siberians were about to explore was a land unoccupied by humans but far from empty. Here roamed the most impressive array of wildlife ever encountered on earth by hominid explorers. The Late Pleistocene tundra teemed with animals. Giant herbivores and huge carnivores wandered or prowled the Arctic steppes. For the earliest explorers, it must have been the wildest landscape on earth. The routes were unknown, the terrain and river crossings treacherous, the country haunted by huge beasts that could kill and eat you.

Some authorities, not many of them archaeologists, think the continent was uninhabitable until some of the giant predators died off, notably the huge, pack-hunting lions, the sabertooths, maybe the cheetah and certainly the gigantic short-faced bear. Could the threat of becoming dinner and the difficulty of securing kills constitute sufficient determent to seriously curtail human colonization of the lower states? Here's a consideration I will chip away at throughout this book.

∼

Before my love of this epic story carries me away, I should mention that there is precious little archaeology from the Late Pleistocene to provide details or dates for the early days of the Great Adventure. These few dates, however, tighten the noose of plausibility and provide loose parameters for the telling of this story.

The oldest Siberian sites close to the Bering Strait date to less than 14,000 years ago, and there are only a couple such records. Otherwise, only a Siberian find called the Yana Rhinoceros Horn Site in far western Beringia dates back as old as 30,000 years ago. There's no archaeological record at all in northern Siberia between 30,000 and 14,000 years ago—the same for Alaska. The negative evidence may not mean much: Northeastern Siberia is even more remote than the Alaskan tundra.

In North America, a small handful of archaeological sites, most of which are perched along tributaries of the Yukon River, date around 13,300 years old. Very old dates, 50,000 to 30,000 years ago from South Carolina, Mexico and South America, are mentioned in the literature but are widely criticized as unlikely or outright loony. A single site way down south in Chile has yielded dates from radiocarbon samples, contested by a few prominent archaeologists, of at least 14,000 years old. Other dates older than Clovis (called "pre-Clovis") are claimed for Pennsylvania, Wisconsin, Washington, Oregon, Texas and the southeastern United States. Until Clovis people showed up 13,000 years ago, that's about it: Very little evidence of human presence and the oldest solid date lies at the opposite end of the American continents from the presumed port of entry at the Bering Strait.

With these dates, archaeology provides at least a hint of what life was like in the last days of the Ice Age, but we are left to try to fill in the details, dangers and daring of everyday life.

Genetic studies, often ciphered in an unfathomable idiom, suggest humans slipped into Alaska from Siberia around 30,000 years ago and, in the absence of information from archaeology, constitute perhaps the strongest argument so far for very early people in the Far North. Linguists want even more time, 36,000 years, to account for the divergence of language groups in America. People of the North, if they were indeed up there, would likely want to go south, where their oral histories from the Old World told them the living was easier. Also, the flocks of waterfowl they depended upon in summer flew south for the winter. The great glaciers of the last ice-age interfered with the people's migratory abilities.

The last big run of ice in North America climaxed 19,000 or 20,000 years ago, when the Laurentide ice sheet spread west and collided with Cordilleran glaciers of the Western mountains. Glaciologists call this event the last glacial maximum and it marks a time when most scientists think ice would have blocked all routes from Beringia down to the lower states south of the ice. The LGM is an important date in American prehistory because it blocks out the time periods and routes by which the first Americans could have colonized all lands to the south of the glaciers.

~

Three routes—three main theories (there are also trans-Atlantic and Pacific hypotheses)—have been proposed for reaching the land south of the ice, to mid-America from Beringia: First, walking down before the LGM; secondly, using boats to navigate the Pacific coast after the LGM, about 14,500 years ago; and, last, using the ice-free corridor from the Yukon down along the Rocky Mountain Front sometime after 13,300 years ago.

The earliest possibility for a terrestrial way south is people walking down from Beringia before the LGM and the glacial sheets covered all routes. Archaeologists are not big on this route because there's no firm evidence of humans in the Americas before about 14,000 years ago. The Pleistocene Beringians, if they indeed colonized the New World before the LGM, seem to have vanished into the Arctic mist until it was time to go south. So there's no compelling reason—no data to explain— for professionals to work on this theory. The pre-LGM route meant walking down the glacial-free valleys at the eastern foot of the Canadian Cordillera between 30,000 and about 20,000 years ago. Some glaciologists think there were other routes south before about 21,000 years ago or whenever the colliding glaciers slammed the door shut. The contested key to all these routes is knowing where the ice was and when it melted. In any case, if humans occupied Alaska and eastern Beringia during the time of an open route, they could have made it down with relative ease (perhaps excepting the considerable threat of giant predators, especially the short-faced bear). How do we know this? Because a brown bear ambled down from Beringia by at least 26,000 years ago (how much prior to this time is unknown). Any route a grizzly bear could travel, so could human beings. The grizzly fossil was found near Edmonton, Alberta—not far from the southern region of the ice-free corridor.

What archaeological evidence would lend credence to this proposal? Any substantiated date from the lower 48, or perhaps Central or South America, which proves older than the time the coastal route became passable, around 14,500 years ago. Archaeologists have found sites in Pennsylvania and the southeastern United States with dates this old (roughly 16,000 to 19,000 years ago). Those radiocarbon dates are disputed but if any pass muster (Pennsylvania is a good candidate), we need to think about people surviving with giant predators and why we have yet to find considerable evidence of humans or of their passage prior to Clovis.

Boats, the eternal wild card, could have arrived—dodging icebergs in a terrible sea—from the Pacific Rim or Atlantic ice at any time in the past 40,000 years.

The second possibility, the coastal route, also unsupported by evidence, sounds more promising: Humans could have come down the partially de-glaciated Northwest coast about 14,500 years ago, paddling watercraft around calving glaciers and living off shellfish and other marine resources. Advocates, and there are many, for the coastal-entry theory point out that grizzly fossils dating back more than 13,500 years have been recovered from islands off southwestern Alaska, indicating a refuge from the ice. Recent sediment cores from Aleutian Alaska indicate the deglaciation could have started even earlier. A few archaeologists suggest that Clovis technology evolved from the hypothetical marine-mammal hunting tools of people who could have paddled down the West coast. Smaller arguments storm over whether the first humans walked down the coast, paddled boats or partook of a maritime economy. Some authorities think the Americans would have walked down from Alaska because the sea was lower by as much as 360 feet and the route 14,000 years ago would not have been the challenging, impassible coastline of today.

But this still seems forbidding. Fourteen thousand years ago, the exposed continental shelf would have been carved into wild canyons by raging rivers fed by glacial melting. The now-drowned coastline would have been at least as difficult as today's impossible-to-walk coast. The amount of ice calving off into the ocean 14,000 years ago was even greater than today's flow. The retreating ice created huge fjords as the ocean flooded the huge trenches gnawed into the bedrock by the retreating glacier, miles-wide inlets and glacial heads. That greater volume of glacial melt still had to flow into the Pacific via countless forbidding rivers. And walking from Beringia to the lower states on a trail of broken mussel, clam and oyster shells, past salmon rivers and tide pools teeming with seafood: Could anyone refuse a maritime dinner? This material is discussed in Chapter 5.

The task for archaeologists trying to document the coastal route remains to find sites that predate Clovis—underwater, in dry caves or upon isostatically rebounded (from the removed weight of the melted glaciers) headlands.

The last route, migrating down the ice-free corridor (IFC), was proposed fifty years ago. The theory was a component of the "Clovis First" proposal whereby early Alaskans living along a tributary of the Yukon River dashed south down the barren landscape of the corridor, subsisting off what foods they could carry and migratory birds about 13,100 years ago or slightly before. Here, anthropology presents a hypothetical account as opposed to a rigorous model.

This theory speculates that these Clovis progenitors would have first encountered mammoth near the southern end of the corridor, on their way southward, sized them up for dinner and wondered how they might bring down one of these big animals. After all, the story goes, just centuries before, their great great-grandfathers hunted mammoth and they had all heard the stories told around the campfires of delectable mammoth feasts. They wanted to kill one to eat but their spears were too puny. Just south of the northern loop of the Missouri River, the travelers stumbled across big lithic quarries of quality knapping material where they could experiment with their stone-flaking techniques. Mammoth-necessity inspired function and these migrants invented the iconic Clovis projectile point. Fanning out from the Rocky Mountain Front, the theory suggests, they pursued their prey and maybe hunted the megafauna into extinction. Clovis technology spread throughout lower North America, within 200-300 years. The continental U.S. might have been uninhabited at this time or sparsely populated with humans co-existing with the giant predators in the mid-latitudes.

Parts of this hypothesis hold up and others don't. Archaeological sites from the corridor pre-dating Clovis have not been found. Clovis people were not first (earlier dates come from Chile, Oregon, Wisconsin, Pennsylvania, the southeast U. S and probably Washington and Texas) nor did they eat up those great big animals all by themselves, though they no doubt helped finish off the last mammoths. But the dates for the final opening of the IFC and the appearance of the Clovis people are uncannily close. And, as an American colonizing event, Clovis eclipses all previous human probes by light-years.

A number of big questions for archaeologists remain about Clovis culture, the first widely recognized archaeological presence in North America, big-game hunters presumably specializing in mammoth. The near-synchronous appearance of its magnificent signature artifact—a large, fluted, exquisitely flaked projectile point usually crafted from the finest cryptocrystalline rock sources—across the country from Washington State to New England, to Florida and Panama within a few hundred years, is considered one of the most amazing events in human colonization. Where did this technology come from? A remaining question is whether Clovis technology exploded across an unoccupied land, or did it diffuse across a preexisting population?

The opposition to the "Clovis First" hypothesis has been in part reactionary, as some archaeologists felt they had had this theory crammed down their throats for thirty years. Thus, the "Clovis First" idea has been attacked with scorn, ridiculed and all but dismissed by popular science and many academics. The dismissive phrase, "the final nail in the coffin of Clovis First," shows up many dozens of times in the archaeological literature. There are reasons: Several widely accepted radiocarbon dates place humans south of the ice before the appearance of the Clovis people.

So, what route did early humans use to get around the ice? Probably, they used all three. Whether they survived or thrived is another question.

~

Additionally, there are a few trans-oceanic migration notions, considered fringe theories by most, but not all, scientists. These propose that the First Americans drifted or paddled over from South Asia, Australia or—postulated by well-established archaeologists—the Iberian Peninsula of Spain and France. The lithic technology of the Solutrean tradition (people who lived in southwestern Europe about 22,000 to 18,000 years ago), they argue, is the true progenitor of the Clovis point and this very terrestrial-adapted European culture, with no evidence of maritime technology, overcame a very cold ocean over a time span of 5,000 years by iceberg-hopping in skin boats, presumably living off sea mammals in order to deliver the distinctive Clovis weapon system to the Southeastern United States. A bothersome insinuation of the primacy accorded to European lithic diffusion (technology passed on from one people to another) of the "Solutrean" theory is that Native Americans couldn't somehow have invented the Clovis point on their own. These are troubled waters. Proponents of the European origin of Clovis claim the intentional over-shot ("*outré passé*") flaking technique, whereby a flake is struck with such force that it continues over the top of a biface to the other edge, is unique to Clovis and the Solutreans.

This is a very rough archaeological map of the debates raging over the peopling of the Americas.

~

Archaeological presumptions underlying much of the conjecture surrounding routes are founded on negative evidence. As in the case of Beringia, where sign of humans appears to vanish for 15,000 years, the nuanced interpretation of finding nothing is important. Absence of evidence is not, they say, evidence of absence. Having tracked Siberian wildlife, I know that this Russian piece of Beringia is an exceedingly remote place. Northeast Siberia resembles the wild interior of Alaska without the history of gold mining. Not finding evidence of humans in the fertile valleys of America's Southeast, East coast and Midwest (more Clovis projectile points, mostly surface finds, come from this area than anywhere else), where farming and modern development have exposed the deeper sediments, carries far different implications than scarcity of sites on the empty tundra and muskegs of Siberia.

The lack of human evidence in Arctic Siberia, and across Beringia, from about 30,000 to 14,000 years ago, used by archaeologists to show that no humans lived there, has an analogous argument from paleontology. The fossil record of brown bears from Alaska shows a gap from 35,000 to 21,000 years ago. To some paleontologists, this means the grizzlies weren't there during that time span and that maybe the short-faced bears drove them out. Until 2004, the oldest grizzly fossil in the lower 48 was thought to be about 13,000 years old. Then a grizzly skull dating 26,000 (radiocarbon) years old showed up in south-central Alberta. Twenty-six to thirteen thousand years is a big gap in the record—at least 13,000 years—and I've heard no claims that the brown bear vanished from the areas south of the ice during this time.

The ease or difficulty of locating archaeological sites depends as much on the density of ordinary human activities as it does field surveys by professionals. Siberia and Alaska are tough archaeological nuts to crack, as is—despite surveys—the ice-free corridor where a small number of people may have passed so swiftly they left no easy markers. The modern northern Rocky Mountain Front is sparsely populated, very sparsely in key locations like the southern end of the corridor, the Plains are plowed and most of the Midwest and East farmed, developed and very accessible to arrowhead hunters. The point is that negative evidence should be weighed with caution and shaded with horse sense not cherry-picked.

There's no direct archaeological data to support any of these migration theories, no sites older than Clovis in the corridor or on the Northwest coast, no progenitors of the classic Clovis point that could have been invented in America or come from Spain or from maritime hunters trekking inland. Indeed, we probably shouldn't expect to find sign of the passage of a few, swift explorers who might not have left much non-perishable evidence behind.

Radiocarbon dates older than 23,000 (pre-LGM) years old from the lower 48 or South America are routinely dismissed as misinterpreted, contaminated or crazy. The three routes are default positions: if people couldn't get down the corridor, they must have come down the coast and so on.

Here is a field where entire professional careers, major theories, fortunes and fates may turn on a single scrap of dateable refuse from an ancient fire pit. One need only mention that scientists are looking for now submerged arrowheads in the Pacific Ocean in order to conjure up the coastal route and at the same time totally debunk or otherwise discredit advocates for the corridor theory—without offering much in the way of evidence. Generosity and camaraderie are not hallmarks of this discussion. The *Washington Times* once described North American archaeology as "one of the nastier academic communities on the planet."

This is not to say that early Americans did not come down the West Coast before Clovis times. They likely did. There just isn't any archaeological evidence to prove it yet.

∼

Any child who has picked up an arrowhead and then wondered about the lives of the people who left it there knows the magic and excitement that fortifies the academic field of archaeology. It is perhaps the most followed profession of our popular social sciences, filling the pages of *National Geographic* as well as *The New Yorker* and occupying many hours of cable television. Yet the discipline of archaeology appears to linger on the cusp of academic respectability. This could be a holdover from the profession's 17th or 18th century origins spent looting Mediterranean tombs or digging up thousands of Native American burials a couple hundred years later. Even the most famous of today's professionals may mull over the founding question: Can the field of archaeology ever be pursued as a science?

Even when a valid breakthrough discovery is made, say a solid pre-Clovis date or a site that challenges conventions regarding Clovis-blitzkrieg or Pleistocene extinction, those claims must be pitched to the popular press in a scramble for publicity. The only way to keep working on a site or find is by securing funding and volunteer labor (better yet, let the volunteers pay for the privilege) for which there is intense competition. The lead archaeologist must stage a press conference (an outsider might wonder if the field of archaeology would appear more professional if it didn't hold press conferences), announce his theory and then show his goods. Better still, one could sell the site or idea to Nova, the National Geographic Society or the Discovery channel.

This is unfortunate and unfair to the archaeologist, as the amount of the grant necessary to continue digging a site is peanuts compared to other sciences—a few hundred thousand dollars per dig versus the vast millions we dump into space projects and war toys.

We shouldn't ask professionals to be pitchmen for their projects, to hype the importance of finds and sites by claiming the secrets of American archaeology can be unlocked by a single date or a handful of artifacts that look like they came from someone's driveway. Still, to keep the investigations going, the diggers need grants and free laborers. Today's smaller institutions have trouble coming up with the cash. With our current government's support of scientific and intellectual inquiry headed for the Dark Ages, things don't look good. But what is needed is public funding for worthy archaeological projects. For the price of a single stealth fighter jet, we could fund most all of them for years.

Modern excavation techniques themselves are remarkable, far beyond the shovel-and-trowel work I was taught on my first professional dig fifty-five years ago. Everything is saved, logged and measured with lasers. Forensic-like sciences preserve and analyze everything organic. Modern genetics have opened another book on human origins and today's digs sometimes resemble hospital operating rooms with gloves, masks and sterile techniques. Watching field workers devote an entire summer to excavate a few inches of a five-by-five foot grid is a thing of beauty. Despite the publicity that a new site brings to an institution, I'd love to see these analytical tools used to re-examine important older sites, mining the data, physical landscape and museum specimens for new insights.

Archaeology embraces a great deal of hard numbers from the physical sciences as well as the fields of genetics and linguistics. These latter tools are critical for inferring relationships or constructing hypotheses but languages and genes can't be directly dated. What archaeologists want in the end is archaeological data, datable material that puts people in a place at a particular time. Scientists, including archaeologists, also prefer "replicability," which means more than one example, site or date. The problem in the study of early Americans is that very few pre-Clovis sites have been found. Either such sites are indeed quantitatively rare or these early Paleoindians didn't leave behind distinguishable material culture, such as Clovis spear points. Maybe they made perishable artifacts out of fibrous plants, wood or bone and used cobblestones to break open big bones or shellfish. Organic material doesn't get preserved except under rare, anaerobic conditions such as peat bogs. It's probably a matter of both: scant evidence left by very few people spanning two huge continents.

Along with the extreme scarcity of pre-Clovis evidence is the absence of human skeletal remains. There is the elaborate Clovis burial in Montana and parts of another Clovis-age skeleton, not associated with cultural remains, from the Channel Islands off southern California, a woman (oops, recently revealed to be a man) who dated almost 13,000 years old and who required a boat to get out there. That's it. A very few skeletons have been located that are younger by a few thousand years, including the much touted Kennewick Man, of the Columbia River, who died a mere 11,000 years ago.

In the absence of Late Pleistocene human remains, American archaeologists must make decisions about the authenticity of artifacts, datable stratigraphic contexts, and bolster these claims with inferences from geologic and paleoenvironmental studies. Then they need to be ready to defend their conclusions from skeptics in opposing schools.

The field of archaeology has also received the unsolicited and unintended job of gatekeeper at the fortress of Native origin mythologies. The onerous part of this responsibility throws the "science of archaeology," with the limited tools of the profession, into the mud bath of ancient Native repatriations, such as transpired with the remains of Kennewick Man. The arguments of cultural affiliation are never entirely scientific. For an open-minded archaeologist, however, here is an intriguing and exciting opportunity: To contribute to a deeper understanding of where we came from. Even fragmented origin stories are a glorious improvement on traditional European-based written history.

In the wake of the squabble over Kennewick Man, we were reminded that archaeology lingers yet as a barely disguised insult to many Native Americans. The maneuvering and sequestering behind the discovery of Kennewick Man was a media circus fed by assertions of white supremacists that the Aryan race discovered America. Whatever the intent of the anthropologists, the result was increasing polarization between the scientists and the local Umatilla people. In the past couple of years, prominent archaeologists have made end runs on the Native American Graves Preservation and Repatriation Act (NAGPRA) in order to get DNA samples out of the Montana Clovis child burial with no attempt whatsoever to contact local Native American tribes. Here is an issue I clearly do not comprehend and about which I might be totally wrong: Why has mainstream archaeology appeared to stand so firmly against the broader sentiments and spirit of NAGPRA? What's the real threat?

But these examples are selective citations, which a spectrum of the current generation of archaeologists doesn't buy into. Twenty years ago, I co-founded an organization that works with Native communities to conserve wilderness homelands, big ones in the range of 4 or 5 Yellowstone Parks (8-10 million acres). That successful effort (now approaching 25 million acres) has made me optimistic. An especially useful bridge between Native North American and European cultures is using the natural world as a means of finding wisdom, the traditional use of the wild, as valid today as 13,000 years ago.

Might the modern mind contain an old worldview that floats above culture?

∼

None of these qualifications have diminished the huge popularity of early-American archaeology. It is a very attractive occupation, partly because modern excavations often take place in remote, even exotic places where the topography has remained unchanged since the retreat of the glaciers. The fieldwork is conducted in the outdoors, in a landscape that encourages us to imagine how ancient people might have lived, to look for the parallels spanning the ages. The field can be at once romantic yet mundane, probing the dirt with a delicious childishness, lunging towards discovery. A few professionals might resist such a representation, but the appeal of Paleoindian archaeology is a draw toward adventure.

This is not a frivolous notion of adventure. Unlike extreme snowboarding or recreational skydiving, this subject, the surge of the first people into the Americas during a time of global warming, lies close to the reoccurring, central themes of human evolution—bold migrations and technological adaptation to climate change. Life and death decisions were a piece of the daily menu in the American Late Pleistocene. The perception of risk was palpable around every bend of the creek. I can imagine a great concentration of selection occurring during this brief period and the part of the organism it would have most influenced is the human mind. In this tiny fragment of time, with our clumsy tools of analysis, we will of course find no record of such evolution. Yet in piecing together this story—and this is my own interest—by anthropology or raw speculation, we look in at ourselves, at what is possible and what we are still capable of.

∼

It has been my purpose in exploring the earliest colonization of the Americas—a story constructed of interpreted scientific investigations and reconstructed tales of adventure—to ask questions that appear relevant to the 21st century—an effort to draw the Pleistocene past into present day climate change at every appropriate twist of the trail.

I believe in the value of wilderness and it is that wildness which bridges these two worlds. The greatest wilderness ever glimpsed by humans was the uninhabited Americas at the time of first entry into the New World. We all are children of the Pleistocene: Will we dare face the hot future with the ballast of those pilgrims who charged out of the Ice Age?

An important reason this ancient story resonates well today is that it was lived by people much like ourselves who braved New World habitats whose considerable remnants are still with us. They are the familiar landscapes of Alaska and Canada as well as the American West. It's the same American wilderness I have been writing about for four decades. And though the mammoths and sabertooth cats are gone, grizzlies, musk ox, caribou and bison are still around. The post-glacial landforms are also unchanged and those ice-age plants still grow here arranged in slightly different vegetative communities. It's a life a few of us still live and all of us can imagine. We may not know what language these first Americans spoke but we know by inference they were bands of big game hunters, families traveling through this brand new land. We could, if we choose, comprehend much of what these people went through. There are scores of scientific papers describing the climate, the geology, the flora and fauna of the landscape through which these ice-age explorers may have traveled. We know something about their technology, what tools they carried. What is lacking is a more intimate picture that vitalizes this breathtaking landscape, a sense of what it felt for individual humans to face the crossing of a raging river, a glacier, hunt a mammoth or co-exist with huge predators.

A practical bridge to the lives of these people, wondering what it must have been like 13,000 years ago, is comparing the perils of that era with the challenge of today's wilderness. It's a reasonable way of thinking since many of the habitats and most of the landforms are the same. Contemporary archaeology realizes the value of this comparison but in general lacks the experience; it offers up the image of buzzing tundra grizzly bears with a helicopter to conjure the wild (it's unfair to pluck out a single example but it's very close to my own home). Having spent much of my adult life on foot with grizzlies, I believe we can improve on this approach. One small edge I have as a chronicler of this tale is my own experience in the wild: Five decades of camping out and traveling the same routes on foot as these ancient pioneers, tracking grizzlies, polar bears, wolves, a few jaguars or tigers and foraging off the land as much as practical in all these habitats. As a life-long advocate for wildlife, I have, in particular, spent a lot of time tracking and living with brown bears throughout North America. The places grizzlies still live are the wilderness regions of Alaska, the Yukon, British Columbia and the Rocky Mountains all the way down into Mexico (I saw sign in Chihuahua in 1985). This vast area is exactly where the story of the First Americans unfolds.

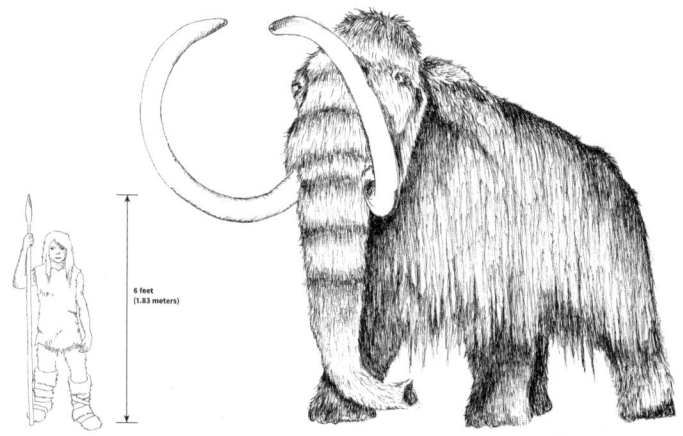

6 feet
(1.83 meters)

I intend to bring some of the lore of the land into the telling of this story, infusing natural history into speculative scenes of early American life. I'll inject a few stories from my field-notebooks, intended to illustrate the difficulty, or ease, of land navigation and resource use. We'll see how they wash. This material is aimed at a general audience from the viewpoint of an informed outsider and hopefully will ground the older tale of the Pleistocene in a familiar natural setting we can all recognize. If modern Paleoindian archaeology is to be of any value to understanding and facing 21st century climate change, we must somehow connect the literal terrain. The key, I believe, lies in the original landscape that survives as today's wilderness.

CHAPTER 4

Invisible People

Ice-age People in the Far North
of Siberia and America

HERE IS A FASCINATING AREA of inquiry that has received very little attention: Humans surviving in Arctic Siberia and Alaska before the LGM, roughly the time between 15,000 and 30,000 years ago. During this period, only a single unchallenged site has been located in this vast region, the well-dated (27,000 radiocarbon years ago) Yana Rhinoceros Horn Site in far western Beringia. But that's it. Nothing else credible has been found older than about 14,000 years old, in either Siberia or Alaska. The pervading view is that these well-adapted Arctic pioneers simply retreated south to central Siberia until the climate warmed after 15,000 years ago.

But there is another perspective: That we have yet to find the evidence that would document early colonizers in Siberia and Alaska. The Yana site is characterized by frost-heaves of blocks of frozen sediment containing fossils of extinct animals in association with human artifacts. With global warming, today's Arctic permafrost is rapidly thawing; solifluction and other geologic processes may reveal undiscovered artifacts, encouraging fresh archaeological investigations of the Far North. We may find sign of these invisible people yet.

$\approx$

So what's the larger picture of human colonization of the Americas? Modern man and woman evolved in Africa as early as 196,000 years back and started roaming other regions of the world by at least 65,000 years ago. By 40,000 years ago, humans had penetrated nearly all habitats of every continent with the exception of South America, Arctica and North America. Why not the Americas? Maybe *Homo sapiens* had trouble getting here: The surrounding seas presented a formidable deterrent and the frigid Arctic made expeditions to the Bering Strait daunting. Or if they did get to America, perhaps the fierce predators of the Late Pleistocene turned them back. Early probes by Asian or European explorers could have disappeared, leaving no record of their passage. They could have drowned, died off or been eaten. Why did it take them so long to get here?

There are a number of scenarios and hypotheses for the earliest visits to the Americas. Some of these possibilities are supported by indirect evidence. Others are mostly speculative.

The presumption here is that the first Americans were modern humans, not Neanderthal or other species of hominids. And that assumption could be wrong (but most authorities doubt it): 40,000 years ago, in the European eastern Arctic, at the northern end of the Ural Mountains, someone left tools and an incised mammoth tusk. These people could have been Neanderthal.

≈

Australia was colonized by modern humans a number of times, perhaps beginning as early as 65,000 years ago. These migrants would have come out of Africa, probably escaping a widespread drought about 70,000 years ago, about the same time as Mount Toba in Sumatra blew— the most powerful volcanic eruption hominids have ever experienced. These people bypassed the Neanderthal occupation of Europe, slipped eastward and colonized southern Asia, then island-hopped south to Australia. That journey south took them over 55-mile stretches of open ocean, out of sight of land, and required sophisticated marine navigating capabilities. Fishing the shallow seas—assuming the use of nets—would have been a breeze.

That means people also had boats throughout Indonesia, South and Southeast Asia. Modern humans lived on the northwestern Pacific Rim. Human skeletal remains, dated 35,000 years old, were found in the Ryukyu island chain of Japan and indicate the use of boats. Would these sea-faring people have sat sunning themselves on the beaches of the East China Sea for 40,000 years without wanting to explore the rest of the Pacific Rim? Maybe Asian sailors hugged the northern Pacific coast or came across the Aleutians to take a look at the Americas. If so, what did they see: a bunch of huge cats, big wolves and gigantic bears, among the other large critters? Maybe these Pacific circumnavigators didn't attempt a serious landing until they arrived in South America, south of the most dangerous predators. Maybe they survived, at least for a while, in small enclaves in South America. Some very old, and sharply contested, archaeological dates have been reported from South America.

Controversy entangles much of this material but one interesting suggestion by scholars is that the cognitive ability to come up with a boat/net-fishing technology 50,000 years ago is not unlike that of figuring out how to live in the frigid North where fire, clothing, shelter and storage are essential. At any rate, 40,000 years ago modern humans around the world were adapting their lives and technologies to marine, Arctic and coastal environments.

Given such ancient mariners, the Bering Strait land bridge would not have been necessary. Such boat people could have hugged the coast or, more dangerously, they could have slid across from the Kamchatka Peninsula hopping across the Aleutian island chain. Sea birds would have fed them as well as shown them the way across the bigger stretches of open sea.

Glacial cycles create rising or falling ocean levels over continental shelves. At a midpoint, like tidal pulses, shallow seas are created and these seas are biologically very productive of nutrients. The presence of a shallow sea would be a very good place for boat people to go exploring and corresponds generally to the times of settlement suggested by North American geneticists. Crossing the wide Pacific, at any time, would be a dangerous stretch of mind and sea.

Forty thousand years ago, incidentally, seems to have been a time when modern people were on the move, a warmer time of glacial interstadials and productive coastal habitats. Hordes of prototypical modern humans were pushing into Neanderthal-occupied Western Europe and also spreading eastward into continental Asia.

Some authorities think these seafaring people would have probed North America, trying to slip in, numerous times during the past 40,000 years. Except for a few dates of 50,000 and 33,000 years ago, which very few accept, there's no American archaeological evidence of these expeditions—it could be underwater. But what kept them out? Pleistocene predators: a pride of huge lions, bears, sabertooth cats and wolves charging the boat on the beach at Malibu?

Exploring inland in South America, as some suggest, might not have been such a suicidal probe for these ancient mariners with fewer great predators, namely the stouter but perhaps slower sabertooth *Smilodon*, on the prowl. That date of 33,000 years ago, incidentally, is from the Chilean site of Monte Verde and even the primary investigators aren't certain the date is archaeologically solid. But if early mariners probed the Pacific Coast of the New World numerous times and were disinclined to land in North America because of zoological dangers, the sailors could have pushed on south and made land in South America.

If the 33,000-year-old hearth proves valid, where else would these people have come from (Monte Verde is about 35 miles from the Pacific Ocean)? Maybe Monte Verde represents a pocket of survivors from early boat-people, instead of the result of one of those long slow walks from Beringia postulated by archaeological models. There are other debated early dates from Central and South America: 25,000 years old from the Chapala Basin of Mexico, 16,000 years ago from Pedra Furada in northern South America and a similar date from the Taima-Taima site in Venezuela. Maritime colonization might explain these sites.

Some of the big bad Pleistocene megafauna, *Arctodus simus* in particular, did not make it to South America. Neither did the brown bear: The grizzly ranged down the Sierra Madres far into Durango, southeast of the deserts, not far from the Chapala Basin, and was poised for a colonization of tropical habitats that would have led the species south into the Andes of South America. European firearms put an end to the bear's expansionism.

Arrival in America by boat from the Pacific will remain the eternal hidden possibility. Humans could have landed most any place at any time in the past 50,000 years. The Iberian connection, another well-boosted hypothesis, contends that Europeans from the Iberian Peninsula paddled across the Atlantic 18,000 years ago, bringing their Clovis-like lithic technology with them. Thus the Solutrean culture of Spain and France morphs into Clovis 13,000 years ago in the southeastern United States. Unresolved problems with this theory include that the Solutreans seem to have been landlubbers and had trouble, some authorities think, crossing European rivers; they would have had to overcome 3,000 miles of the icy Atlantic Ocean and then there's a paucity of evidence for the 5,000 years these people hid out in Virginia.

~

An overland migration before the LGM from Siberia across the Bering Strait to the Alaskan side of Beringia remains a possibility, although there's almost no evidence of people in the Far North at this time. They survived for a while on the Siberian side. Some hunters might have hung on in Alaska during the frigid millennia before the glacial sheets spread across the North American continent. Or these speculative ice-age pioneers could have gone south to the lower 48 before the ice slammed the door shut.

We know people lived in the Siberian Arctic 30,000 years ago, only 70 miles from the Arctic Ocean. They could have pushed east and crossed the Bering Strait on ice or land into Alaska. More recent coastal people of the Arctic have been known to make long latitudinal runs along shorelines. If they did, they would have few choices: Stay in eastern Beringia until the coast or an interior corridor opened up or find a way south to mid-America before the ice sheets closed off all routes south during the LGM. These travelers would have to co-exist with lions, sabertooth cats, big wolves and short-faced bears—a formidable task. But maybe some made it south and evaded the predators, especially the gigantic bear, by moving into the forested regions of Pennsylvania or southeastern Wisconsin at a mammoth butchering site that could represent a pocket of survivors from a Pre-LGM inland route down from Beringia.

∽

I wanted to tell the anchoring story of this book chronologically, from the time humans first stepped foot on North America to the last days of the great megafauna and, right off the bat, I stepped into a bog of opinions and presumptions buoyed by very few facts. A huge range of possible dates covers this icy human landscape and very little archaeological evidence has been found to narrow down the guesswork.

Controversy surrounding the time of first entry still reigns. Back in the 1960s, Louis Leakey, of Olduvai Gorge fame, found what he thought were human artifacts in a southern Californian alluvial fan that appeared to date at more than 100,000 years old. A few years ago in South Carolina, a chunk of chert, believed by the excavators to be a human artifact, was dated, by associated charcoal, at 50,000 years ago. A couple of authorities believe human footprints from Mexico are 40,000 years old. Human fingerprints found in a cave in New Mexico were reported to be 37,000 years old.

Few anthropologists buy into anything that old. In fact, most archaeologists studying early Americans won't seriously consider human entry prior to about 14,500 years ago. They have their reasons: Nothing older has been found in the Americas that mainstream archaeology accepts as unimpeachable evidence.

Yet ancient people could have come over earlier. They probably did. But where exactly should the story start? One might be tempted to reach for the science fiction here (some of it's not half bad).

One might think that ethnographic modeling or studies of primitive hunting groups might be useful in reconstructing the unrecorded lives of these first human visitors to eastern Beringia, but applying this approach across two continents has limitations. The early Americans are a special case in a few different ways and there are no real anthropological precedents. First, no people lived in the Americas—thus no demographic pressure, no territorial squabbles. Secondly, the new land was teeming with great beasts, some never before encountered by humans. Usually when we imagine humans encountering never-before-hunted animals, we think of islands, Tasmania, creatures like the dodo on Mauritius, the great auk or the moa of New Zealand—critters we drove into extinction in record time. Late Pleistocene America was too big and new for the limitations of island biogeography theory to make a big dent in a wildlife population that contained such a large number of huge creatures—for a while anyway. Human impact, from a speculative population of unknown size, on native species (Pleistocene extinction) would take millennia to rear its ugly head.

The most important archaeological date framing the earliest possible forays into North America is from the Yana Site in Siberia, what used to be far western Beringia. Though considerably west of Alaska, no insurmountable barriers lay between the Yana people and North America. Why wouldn't these hunters want to explore eastward, along the coast, crossing the frozen rivers in wintertime?

Yana is an intriguing site. The rhino foreshafts are similar to the elk foreshafts found in the Anzick Clovis burial, although separated geographically by thousands of miles and by 16,000 years of time. The lithic assemblage is unlike later Siberian sites: There are no prismatic blades. Tons of mammoth bones are scattered among the cultural materials at the Yana River. How hard was human life in this frozen land? What are the possibilities of living in the far north of western Beringia before the LGM?

> The early spring wind slapped the bison skin covering the opening to the smoky yurt. The winter had been mild, but Siberia was not a forgiving land 30,000 years ago, especially here on this great river, ten degrees of latitude north of the Arctic Circle—the land that would later be called Beringia. Inside the yurt, arranged around the central hearth, sat four generations of family, seven adults and four children. One of the adults, a white-haired elder with a withered arm, was two decades older than the others. Their clothing was sewn from the skins of several animals: Sealskin covered mukluks, caribou shirts and a cape fringed at the neck with wolverine fur. One man flaked butchering tools from a river cobble. Another worked a slice of mammoth long-bone into a sharpened spear point. The women and an older girl pierced animal skins with stout bone awls carved from wolf anklebones, fashioning covers for summer yurts from bison and caribou hides. In another month, they would move the camp out of the sheltered draw of their winter site, up onto a higher terrace where the

Arctic wind kept the insects at bay. During deepest winter, the sun had disappeared altogether for more than two months. Now the warming sun lingered just above the horizon during midday. The days were growing longer. Soon, it would be time to think about the spring hunt. It was almost the season for the river ice to break up.

Outside, the noon sun revealed a white wilderness with no trees. A few river terraces were visible in the numerous channels where the great river began to divide as it approached its delta. The smaller of these distributaries were frozen solid. Only a dark band of water bubbled out of the middle of the big river. A dozen yurts dotted the whiteness. This was a large band but when they left their homeland far to the south, they knew they would probably be on their own for finding mates for their children as well as feeding themselves. Fortunately they discovered a place where animals crossed a dangerous river. Many animals drowned and there was almost always a dead bison or mammoth caught up in the ice flows. (The scene is reminiscent of Lewis and Clark's discovery of the Great Falls of the Missouri in 1805: Fat wolves and grizzly bears lounged on the bank below the falls, feeding on the innumerable carcasses of bison that had drowned crossing the river upstream.) Driftwood lined the sides of the river channels. Here they made their camp. Prowling around the mired and dying mammoth, however, prides of Pleistocene lion looked for a chance at a kill. Wolves skulked just behind the bigger cats and, once the feast began, brown bear and wolverines frequented the carcasses. Humans had to wait in line or try to force their way in to get a share of the meat. Sometimes, the hunters found a stranded reindeer, bison, horse or mammoth to finish off by themselves. They hurled bone spears mounted on foreshafts and stabbed with sharpened sticks until the animals died. If the wounded game escaped, a pride of lions would usually get it. But the lions liked to hunt for live meat and didn't scavenge much. They didn't compete with the people as much as the bears and wolves did.

One summer, the band sighted a rare creature caught up in the ice flows: A woolly rhinoceros struggled along the riverbank. The hunters grabbed their spears and took off running downstream. The two-horned creature pawed at the bank with his front hoofs; huge blocks of ice knocked him back into the current. The rhinoceros lunged again and finally pulled his hindquarters up onto the grassy bank. The animal just stood there, apparently exhausted from his ordeal. The hunters were wary of this fierce beast. The woolly rhinoceros of the tundra was a solitary creature that the hunters seldom encountered. When they did, the near-sighted horned monster would charge blindly at them. Once, a rhino caught a seasoned hunter by his parka, threw him to the ground and crushed the hunter's arm. The man, now old, survived. The woolly rhinoceros, like the mammoth, had become a totem of the people, a beast both feared and revered.

The hunters carefully closed in on the stationary animal that appeared to be offering himself up as meat for the band. Four hunters hurled their spears at one side of the rhino, and then dashed away to reload with another foreshaft. On the opposite side of the beast, three men drove fire-hardened wooden spears into the rhino's body. The animal spun around and took another volley of spears. The hunters retreated to a higher terrace and waited. They had mortally wounded their totem animal.

The entire village—two clans totaling nearly seventy people—gathered around the dead rhino. They built fires to keep the bears and lions away. The old man with the crumpled arm came forward and circled the kill, sprinkling red ochre on it with his good hand and clutching a braid of burning grass with the feeble one. The butchering began. The people used choppers, sharpened flakes and scrapers to remove all flesh from the bone. Two older hunters, probably shaman, carefully removed the big and little horns from the rhino carcass with stone tools they pulled from leather pouches, scrapers and bifaces flaked from quartz crystal. The power of the animal resided in its horns.

The days grew longer. Blocks of ice bobbed down the wide river. The herds of horses and bison returned, looking for a way across. Almost every day, people saw mammoth. The big lions were also on the scene, so hunters approached their prey with caution. But the band craved fresh meat. Much of their winter food was drawn from meat frozen and cached from the fall hunt. Wolverines and foxes sometimes raided the food caches but brown bear hibernated during the coldest months. The people could deal with wolves, wolverines and foxes though a big bear was a more challenging problem. Most importantly, the birds were back.

The band moved their yurts to higher ground, but still out of sight of the prey animals who moved around the camp. Women went to work weaving fresh nets in preparation for the harvest of migratory waterfowl. Hundreds of thousands of ducks, geese, swans and cranes swept up from the south to nest on the muskegs and tundra ponds. The men posted lookouts for hunting opportunities as well as prowling lions and worked on their tool kits, shaping long, beveled rods from the larger horn of the rhinoceros. Other men carved ivory foreshafts from sections of mammoth tusk.

Waterfowl constituted a crucial summer food for the band. Some months, birds made up the bulk of the diet. Women and older children did most of the hunting. Strategies for hunting waterfowl came in many forms but the most efficient way to catch the bigger cranes and swans involved the use of nets.

Daylight lasted nearly twenty hours a day now. The sounds of geese honking and the rattled trumpeting of cranes became deafening. Nets could be strewn about the ponds and kettles at night and, by morning, would produce a few entangled waterfowl. But, more efficiently, lines of nets on poles were placed in the ponds and, under cover of the brief spring night, the women would steal into the waterways and hide by the poles until daybreak when the others would conduct a drive, frightening the birds to rise and skim the lake, flying into the nets the hunters had lifted into the air. The women and children could supply the band with most of their meat—migratory birds and an occasional Pleistocene hare—during summertime. The men loafed and waited for a mammoth or bison to get caught up in the icy river. Until one autumn day: The cries of women and screams of children startled the men at camp. Someone was in trouble. The men grabbed their weapons and surged out of camp towards the screams.

A Pleistocene lioness had snagged a child and carried her off down river. The men came running with their spears, filing down the dry channel where the huge predator was last seen. Except for the birds, it was eerily quiet.

The earliest opportunity for early Americans living in Beringia to reach the area of the lower United States would have been prior to the Last Glacial Maximum (LGM), before the ice sheets slammed the door shut on routes southward. Such potential routes are sometimes called Pre-Max, or Pre-LGM. The prerequisite for the use of this passage is people actually living in the American Arctic 30,000 to 23,000 years ago. Without a human presence in Beringia, the consideration of a Pre-Max route would be pointless. But again, our inability to find conclusive proof of these fleeting ghosts of the Arctic proves little.

The events for which it would be convenient to have reliable dates are the initial crossing into North America and the times of route availability to penetrate south of the ice. But so far there's no conclusive evidence for most of these events.

The first crossing is an arbitrary event, as humans may have crossed many times, going both directions. Still, it would be useful to know when the first Asian set foot on dry land on the eastern side of the Bering Strait. We don't know when this happened. And, again, ancient seafarers could have visited the Americas at most any time, at most any port in the past 50,000 years.

A few hard dates provide only the roughest of parameters:

The Yana Rhinoceros Horn Site in Siberia with its ivory foreshaft has been dated at 32,000 years old, but that single find is followed by 18,000 years of silence in the archaeological record of northeastern Asia.

Eastern Beringian archaeologists have yet to find unchallenged material significantly older than Clovis times (about 13,300 years ago), though information from other disciplines suggests greater antiquity.

～

Genetic studies of Native Americans suggest an earlier arrival date for people in North America. Early DNA analyses implied an entry date as old as 36,000 years ago.

A more recent DNA study concluded that the First Americans hit Alaska 30,000 years ago, lingered in Beringia for perhaps another 15,000 years, then headed southward. The single crossing, it concluded, was the founding population. This theory, also known as "Beringian Standstill," suggests that during the first 15,000 years there was back-and-forth travel, exchanging genes between peoples in Alaska and Siberia. A cautionary reminder about the ease of travel across the Bering Land Bridge: Recall that, as far as we know, the Siberian woolly rhinoceros and the Alaskan short-faced bear never crossed, which may imply a paucity of habitat requirements in the bridge area for rhinos and scavenging bears or other factors we haven't considered yet. Distributions of gene sequences do not dictate the actual geographical routes that may have connected them.

Around 15,000 years ago, the study suggests, the first people in the Americas hit the road for southern climes and began populating the New World. This southward migration through the Americas, geneticists think, was probably a swift pioneering process. If this thesis were true, the people would have had to use the coast as glaciers closed all other routes at this time. These geneticists conclude from their DNA data that most migration followed the coast, although others suggest that this could represent sampling bias, as Native Americans from other regions of the United States were excluded from the genetic study. There is no archaeological confirmation of this genetic map of North Americans.

I should note that there is nothing sacred about this particular reference. At this writing, a very recent study finds a "rare mitochondrial DNA haplogroup" in modern Native Americans of the Midwest that the authors believe means that the first Americans came from Asia not Europe and entered North America via Beringia through the ice-free corridor 18,000 to 15,000 years ago. The part about the IFC might have been a misstatement as the corridor was probably blocked by ice at that time. Or else the IFC opened earlier than most authorities believe.

Many anthropologists believe there were several waves of migration (previous to the arrival of Na-Dena speakers and Aleut-Eskimos), pulses of colonization, and the date of 30,000 years ago is just a starting point on a ticking genetic clock.

Other estimates of arrival times derived from mitochondrial DNA range from 20,000 to 40,000 years ago. This is complicated scientific terrain, nearly impenetrable for the half-committed layman. Recent books by professional archaeologists do an excellent job of translating genetic studies.

Linguistic modeling infers that about 35,000 years would be required for the divergence of the complex of American languages and dialects. This assumes a singular ancestral language at the time of arrival, which diverged over time at a set rate to come up with the number of Native languages at European contact. Some linguists assert that the diversity of languages in a region reflects the length of time people have lived there. The greatest diversity of Native American languages is clustered along the West Coast, from Tlingit territory in Canada to California.

Studies of pre-Columbian human teeth, like genetics and linguistics, indicate Northeast Asia for an entry point and a "dentochronological" age of arrival at about 15,000 or perhaps 20,000 years ago.

Experts tend to consider these language and tooth clocks to be rougher running than the genetic timepiece, which is itself an informed estimate of the amount of time humans may have lived in North America.

The archaeologists I talk to tend to regard these dates as guidelines, at best, awaiting confirmation by finding human material culture. There is yet no direct archaeological evidence of that occupation, except for the Yana site, which is some 3,700 miles west of the Bering Sea.

Despite this conventional interpretation, let's say, for the sake of discussion, some people were in eastern Beringia before the LGM. Perhaps a few hunters came to America after 30,000 years ago, but not much happened until about 15,000 years ago. The Bølling-Allerød warming 14,700 years ago and the subsequent availability of the coastal route are roughly coincident with the dates provide by genetic studies.

These people would have to had survived, if not thrived, in Alaska and eastern Beringia in order to find a route down south along the unglaciated valleys of the Pacific Northwest Cordillera. Admittedly, it would have been tough to over-winter in Beringia between 27,000 and about 15,000 years ago. Hunting, scavenging and caching meat are essentially the only way to endure the long winter nights in the white wilderness. Maybe a decision to find a route south was made shortly after crossing the Strait, so that the colonizer's time in Beringia was very brief. Some glaciologists think several routes through and around the ice were available. The Pre-Max route is a worthy possibility.

How does this translate for our speculative band of ice-age hunters hunkered down on the tundra?

∽

Twenty-six thousand years ago in eastern Beringia, they would have been on the move. The people had no dogs to haul sleds or travois. Dogs would be domesticated from wolves in China some 11,000 years later. Everyone capable of walking carried a pack. Several hunters fanned out in front of the advancing band, scouting the country for game, for routes and keeping an eye out for dangerous predators. The summer tundra was bountiful with grass for the huge grazing animals.

No trees or woody shrubs grew on the chilly steppes and low-lying tundra. Glacial ice covered the higher mountains. The North Slope of Alaska is flat but people would have had better hunting options up in the breaks of the lower mountains and river bottoms, especially in the interior of Alaska. This is a hunter's habitat. Gatherers would notice only a few fruiting shrubs growing on the slopes in late summer and taproots of alpine pea vine growing on the alluvial benches and river bottoms. The people would see that grizzlies (if the bears were up there and the gap in the fossil record of 35,000 to 21,000 years ago is sampling bias) extensively harvest this food source. But big game was the major objective. Muskeg lowlands made summer travel a slog, but dry ridges and braided rivers were easy going. The people wore form-fitting skin or fur clothing, as appropriate to the seasonal weather, and carried portable shelters composed of birch and willow frame covered by animal hides. They carried big game hunting weapons. Most of all, they knew how to make fire—to keep the frigid air at bay and to ward off the fierce Pleistocene carnivores.

The ability to build a fire at any moment, during all seasons and in all landscapes, was the key to surviving the dry cold tundra of America prior to the last glacial maximum. Some modern ecologists have suggested that the absence of woody shrubs to ignite bone fuel (burnt bone in old fire pits is common in later Beringian sites) might have been a barrier that precluded early settlement (meaning the evidence archaeologists haven't found yet) in Beringia before the LGM—the transition from arid grasslands to shrub vegetation took place about 15,000 years ago.

I'm wary of this projection of selected pollen samples onto the people's adaptive abilities. The band of hunters, if present, could still start fires on these cold grasslands, far from spruce forests and the shrub tundra of dwarf willow and birch.

Twenty-six thousand years ago, the Arctic would have been littered with the scattered skeletal remains of bison, saiga antelope, musk ox, mammoth and other grazers along with the bones of a few of the great carnivores that stalked the herbivores. And there would have been huge piles of dung drying on the ground. Dry dung makes for a long-burning fire, though not an especially hot one. A better fire-starter is dried peat. Peat is partially decomposed, highly organic ancient vegetable material. Peat beds lie all over the Far North and their beds are exposed in the eroded banks of creeks and rivers. Such organic layers are not that difficult to locate and one outcrop of fossil peat could supply enough fuel to theoretically warm a band of hunters all year. They would have to dig it out, dry it and packed off as much as they could haul.

The challenge is carrying sufficient dry fuel on your back in order to quickly build a fire in an emergency: Like a group of big prowling Pleistocene lions, a scimitar cat or a giant short-faced bear coming over the rise to appropriate your camel kill. Such events might have daily problems 26,000 years ago. Your defensive tools are spears, throwing sticks (the earliest evidence of atlatl use is from the Solutreans of Europe) and probably nets and bolos. Most of all, they had social cohesion and fire.

Preparation is necessary: You shred the peat or dung, dry it out near a fire and then stash it in your pack for fire starter. For fuel, you need to dry out large quantities of dung, or peat, designate a few members of the band to carry a full load of fuel and have every adult carry fire-bows, flints, dry shredded peat and enough dried dung to get a blaze going. Once the communal blaze is going, bone and dung can be gathered from the surrounding locale, drying in the heat of the fire, to keep it going.

Pre-Max explorers looking for a way southward, as they would probably be inclined with their oral histories and bird lore, would have had to plan ahead, drying and stashing caches of dung, peat and bone as they traveled. In winter, they would have to store large quantities of combustibles around their winter camps.

Bone burns hot. It's the fat that makes it combustible. A fresh bone burns better than an older one and bones at freshly butchered or scavenged carcasses would be the best. Beringians would collect a pack-able quantity of such material as they roamed the land scouting for game and looking for a way southward through the glaciers.

To get a real fire going in the Arctic, the people would find a ravine out of the wind and ignite the shredded peat with an ember from a fire-bow and drill. They would add the dried dung carefully until they had a small blaze going. Then they'd pile on the greenest, fattest fragments of bones—long-legged waterfowl come to mind—until the skeletal material flamed. An improvised billow would be handy, though a half-dozen children blowing into the fire would do the job too.

I don't think the bold bands of Pre-Max Beringians would have been stopped by lack of a shrub-tundra. Not directly anyway. These hunters could make fires of dung, peat and bone sufficient to cook their food and warm their hearths. The question is whether they could build such fires big and fast enough to keep the now-extinct huge predators and scavengers at bay.

> A quarter-mile ahead, one of the advanced lookouts signals from a ridge with one hand outstretched and flat over his nose and the other arm overhead with the index finger pointed up (fictional, but similar to the grizzly bear signals I use with my traveling companions today), meaning a solitary short-faced bear (as opposed to a family group of mother and young or two adults during the breeding season) is approaching. The band of thirty-some people quickly gathers up the stragglers and looks for defensible terrain. A cliff would be best but they are far from the mountains. A steep ravine leads away from the braided river and heads up into a twenty-foot-high box. It's a tight defile with no turning back, which means a fight, but the people prefer these odds to facing *Arctodus simus* on open ground.
>
> The giant bear mostly feeds on the kills of lions and other predators but he has nothing against meat on the hoof or foot either. He might snag a human traveling solo with his superior speed. Solitary predators like mountain lions, and including some large omnivores such as bears, can't afford to brawl casually as an injury could impede their survival—their ability to kill prey and fight for meat at carcasses—implying the people have a chance against the gigantic bear. These two species are brand new to each other. Until hitting Alaska, neither had seen anything so strange as a standing fifteen-foot-high bear or an upright hominid; the bear has no reason to fear the puny two-legged ones.
>
> The women of the band herd the children into the ravine, almost to the end of the earthen box canyon. The men prepare their weapons and a fire near the mouth of the short ravine, just far enough up the gulch where the short-faced bear can't drop in from the top. The women cut emergency steps in the tundra at the head of the draw where, if the giant bear overruns the men, they might scramble out of the gulch and regroup for a running fight. The two scouts rejoin the group but stay on the high ground where they can see the approaching

bruin and, if the beast insists on going up the ravine after the people, throw great rocks at his head. They gather a pile of boulders.

If the band could hide downwind, the short-face bear might pass them by. It prefers rotten meat and has the best nose in the Pleistocene animal scene, so it will also likely know humans are around. The leaders of the band watch and wait until they are certain the bear knows they are there. The scouts try to draw the attention of the short-face away from the band. The two scouts hunch down near the top of the ravine. *Arctodus simus* doesn't fall for the distractions. Sensing a great number of prey or another appealing smell, perhaps the meat fetor of smoky clothing, he beelines for the mouth of the ravine. Men with spears and bolos greet the bear. The Beringians poke fire-hardened spears at both sides of the great omnivore. A bolo stone strikes the animal near the snout. The bear lumbers forward. The men fall back behind the fire line and stoke the flames.

Now is the moment: two hunters hurl their burning spears at the gigantic bear. The scouts throw great rocks from above, pelting the bruin about the head and back. These wounds are minor, but the short-faced bear retreats from the blazing fire-line, twirls, and ambles back out of the ravine.

The people gather on the river terrace and watch the feared creature swing down the green Pleistocene valley. The bear will find another meal. The band most fears the short-faced bear at the moment they bring down prey, a large herbivore such as a horse or caribou. At such times, only fire keeps the scavenging wolves and bears away until they can strip as much flesh as they can from the carcass and carry the meat away to safe shelters and storage caches. The huge lions are just as dangerous as a bear. The cats hunt in prides and bring down any animal that slows or strays away from the herd. That applies to the band of hunters. They stick together and watch the children carefully.

We know that twenty-six thousand years ago plenty of big animals grazed the tundra grasses in the North and by hunting them people could in theory survive winters. The "Standstill" seen by geneticists also suggests a human presence hanging out in Beringia from 30,000 to 15,000 years ago, though archaeologists scratch their heads as to where the evidence can be found. Gold sluicing operations in Alaska haven't washed ancient human bones from river alluvium. The few dead Beringians would have been eaten by carnivores or buried by their clansmen.

There may not be any genetic trace of these early adventures. They might have died off from disease or been wiped out by giant predators.

The North American Beringian population might have been squeezed through a tight bottleneck. The colder weather setting in about 27,000 years ago must have shocked whatever human population that might have lived in Beringia and prodded them to adaptive action. The alternative strategy for these early American would be to find a route south, down where their elders told them the living was easy and from where the birds flew in spring.

What about routes southward through the ice before the LGM? The earliest possible American Beringians, say 30,000 to 27,000 years ago, might have experienced relatively mild climatic conditions such as humans did at the Yana site in Siberia. But starting about 27,000 years ago, it turned colder though no less productive for the megafauna that thrived on the grasslands. For humans it must have been a challenge to remain in the Far North. Then (after 26,000 years ago) the Laurentide ice sheet began to merge with the Pacific mountain glaciers and corridor routes were closed off roughly 22,000 years ago (there's a spectrum of opinions here). At any rate, that seems to be the range of possible Pre-Max dates—30,000 to 23,000 years ago—for people to have made it southward to mid-latitude North America.

Going south was a definite possibility.

We know it was possible because a grizzly bear did it.

In 2002, a cranial fragment of a brown bear was located in a museum collection of fossils from fluvial gravels near Edmonton, Alberta. This well-preserved bone was subsequently dated, based on two "accelerator radiocarbon dates on collagen," at 26,000 years old (this is the average of two radiocarbon dates; a recalibrated date would be several thousand years later). The salient point is that it is decidedly before the last glacial maximum.

Grizzlies came down from Beringia before the two great ice sheets collided, the Pre-Max route. (We don't know how much earlier. The first grizzly crossed over from Northeast Asia 70,000 to 50,000 years ago. Brown bear demographics will be discussed in Chapter 5.)

So grizzlies roamed Alberta by 26,000 years ago. DNA evidence indicates they arrived from Alaska and that this was the founding population of all present-day brown bear in the contiguous states. These conclusions are deduced from analysis of a single, credible cranial fragment from a gravel pit near Edmonton.

If a grizzly made it down, could humans not have followed? The answer is that if people were up there at that time, of course they could have.

The Pre-Max route could have been the ice-free valleys east of the Cordilleran glaciers or another, earlier passage. Any route a grizzly could use would be cushy for humans. Grizzlies colonize empty lands at the slow rate of approximately twenty miles every five years and the habitat has to be rich. Young male bears often make the first probe into new country but it is the family groups that establish occupation. This means the extension of one female home range—the grizzly daughter often sets up her own small range adjacent to her mother's and reproduces the fifth year. People can slug over barren ground and snow with food in their packs, hunt waterfowl and make the same trip (over a thousand years for the grizzly) in a few years.

Evidence of a grizzly in Alberta 26,000 years ago strongly suggests that a route that humans—if they were in Beringia that long ago—could have used was open for several thousand years and that the habitat along that particular migratory corridor was rich and fully revegetated after prior glaciations.

~

How would the trip have gone, south out of the frigid north down to the lower 48 around 26,000 years ago? How would these early Americans deal with the huge cats and bears? What would they eat? Would they hunt big game or gather wild plants? These explorers coming out of the north had to be primarily hunters in order to survive in the Arctic winter. Now, along the Pre-Max corridor route, they might have had options.

I should clarify again the distinction between an early route to the south used by grizzlies in the Pre-Max (prior to 26,000 years ago) to the later ice-free corridor postulated for the Clovis migration (before 13,000 years ago). The former would have had a passage between glaciated highlands that remained free of ice for millennia, occupied by a full range of Pleistocene plants and animals. The latter corridor had just emerged from the later-occurring ice sheets and was, at first, many scientists believe, inhospitable, inundated by meltwater lakes, barren and raked by frigid katabatic winds. Ice-age people could have traveled either route; the grizzly could not.

∾

Is there any evidence that those ancient Yana River hunters blasted east along the Arctic Ocean, like the Thule-Eskimos a thousand years ago, hit the Mackenzie River delta and turned south? Or stopped and lived in the North? A number of archaeological Pre-Max dates from the North American Arctic have been re-examined in past decades. Virtually all have since been questioned or discarded. But a couple of authorities cling tenaciously to Pre-LGM dates from the Yukon.

Perhaps the most famous such site is the remote Yukon's Blue Fish caves, near the Alaska border, where archaeologists, in 1985, found what they believed was a percussion-flaked tool of mammoth bone dating around 26,000 years old. I stumbled through the area a few years later, rowing a raft down the Porcupine River from Old Crow to the Yukon River.

The country of the Blue Fish is a mosaic of old river channels, muskeg swamps and boreal forests of spruce and fir with a few ghostly cottonwoods scattered along the rivers. The caves lie upriver at the base of a limestone ridge.

I stopped by in summertime, fly-fishing the Blue Fish River for Arctic grayling before continuing downstream through limonite cliffs where dozens of peregrine falcons nested. One day I walked upstream on the alluvial channels of summer, which means easy travel in this land of taiga and muskeg. It's gentle country but the bogs and muskegs make walking the wet flats cumbersome; you try to step on the mossy tussock heads at the risk of breaking an ankle and, when you fall off, you're up to your knees in icy water. Dry ridges make the best walking. Otherwise, use the creeks.

Wintertime is different. A friend of mine, an anthropologist, skied upstream from Fort Yukon on the frozen Porcupine during the dark of winter, on his way to visit the Kutchin people, carrying everything on his back. He camped in the collapsed cubbies of vanished trappers, surrounded by the grinning skulls of long-departed wolverine and lynx peering in at his tiny fires of spruce twigs. This incredibly tough guy makes a good case for Pre-Max migratory possibilities.

The reason I stopped at the Blue Fish, besides wanting to smell the Pleistocene, was that I was trying to feed a few accompanying friends by living off the land and had I fished poorly. It wasn't quite salmon-run time. An endangered whitefish lived in the Porcupine, which I wasn't supposed to catch or eat, but an Athapaskan Indian gave me one. The fish was delicious and I could see why it needed protection from people like me. I could catch northern pike with streamers at most any time, but they were so bony I only used them for preparing ceviche. The grayling were smallish, the big ones barely a foot long, but also very tasty.

Amid the cobbles and pebbles, I picked up what looked like a fragment of fossilized long-bone: Camel, horse? I had no idea: The animal lived beyond my paleontological prowess. I replaced the fossil, having long ago lost any interest in collecting. Incised in the drying mud were tracks of caribou, moose, wolverine, lynx, wolf and a single grizzly. Willows lined the creek and, beyond, a boreal zone of spruce and fir. Higher up, the sparse conifer forest faded into tundra.

Earlier, say 26,000 years ago, this land would have been vast tundra grassland, rich in plant food for mammoth and other Pleistocene megafauna, which over the millennia gave way to birch and then the boreal forest we see today. This site would have been a beautiful place to live.

A few miles upstream, perched below a cliff, lay the dry limestone caves of the Blue Fish where scientists found fossils from horses, bison, caribou and lions. Some bones looked like predators had dragged them into the cave. Others looked like they had the cut-marks from human butchering. Over the years, tools were discovered to suggest the presence of humans dating back 11,000 to 18,000 years ago, maybe older. Archaeologists pondered the scars and fractures on the fossil fragments in Blue Fish and wondered whether people inflicted them or if nonhuman processes could have produced them (the marks on the alleged bone tools could have resulted from predator gnawing or natural geological grinding in glacial or permafrost deposits). There is still no consensus and the credibility of the older dates remains in limbo.

I should add that the claim of human butchering marks on ancient bones is sometimes difficult to accept at face value. Predator gnawing and other natural processes have duplicated butchering marks, based on experiments by anthropologists.

$\sim$

Otherwise, and this includes a site north of Old Crow 35 miles to the northeast of Blue Fish caves (from where there is a contested 40,000-year-old date), there are simply no reliable dates, no butchered bones older than the LGM from Beringia. Although many anthropologists think the climate was impossibly cold and bleak, *Homo sapiens* had survived and lived well with grace and ease in such Arctic environs. There are other lines of reasoning that may indicate that humans could have occupied North America long before a coastal route or an ice-free corridor was available.

What would lend archaeological credibility (besides a site in Beringia) for the Pre-Max theory of humans coming down out of Beringia before the glaciers slammed shut? How about a date from south of the glaciers earlier than 15,000 or 16,000 years ago? That would be the key: A solid date, preferably in North America, earlier than anyone thinks people could have come down the coast. A number of radiocarbon dates, from 16,000 to 33,000 years ago, have emerged from New World archaeological literature. Many of the earlier dates have been debunked and the credibility of all have been questioned. There's a date from Mexico of 23,150 BP and more from Brazil averaging around 17,000 that some take seriously.

There are, however, a few possible candidates from Virginia (about 18,000 years ago) and Pennsylvania (perhaps 19,000 years ago). Additionally, a number of dates from mammoth-butchering sites (Hebior-Schaefer) in Kenosha County, Wisconsin average over 14,000 years old and Wisconsin is pretty far inland from any coastal route. Although no firm consensus on the validity of most of these radiocarbon samples exists, if just a single date or a future one holds up that would mean people reached the lower states before the coast was passable or the ice-free corridor opened.

These most likely sites for Pre-Max entry come from the forested eastern United States, especially Pennsylvania. Short-faced bears, who may have run down and devoured two-legged immigrants or appropriated the hunter's kills, were likely a problem for migrants, especially in the open landscapes of the American West. Maybe these small groups of early pioneers died out.

It would be interesting to look at the Pennsylvania rock shelter, the Wisconsin mammoth-butchering places and the earliest sites from the southeastern United States in terms of co-existing with the short-faced bear. The fossil record of *Arctodus simus* indicates a few bears were on the scene—in certain places like Utah, Kansas, Wyoming and maybe Texas—when Clovis showed up. What about the Pre-LGM East? A short-faced bear specimen from the Saltville Valley of Virginia radiocarbon dated almost 15,000 BP. There are recently recovered fossils of short-faced bears from central Florida. But up north in Pennsylvania and Wisconsin: What is the paleontological record of the particular site, the area or region? Are there short-faced bear fossils? If so, what are the dates from those bones? What's the best guess about how these people lived? Assuming they hunted and foraged, would the huge Pleistocene cats and bears have interfered with the people's livelihood? What, given the terrain and vegetation, would be a strategy for foraging? I tried to look up this information in the literature but didn't get far.

Recently, a 50,000-year-old date on charcoal from a South Carolina site has been reported, but the authenticity of the "tools" found in the same sediment layer has been questioned. Many archaeologists think freezing and thawing or other natural processes created the alleged artifacts.

Thus, very few scientists believe the very early dates—33, 000 to 50,000 years ago—are credible and, even if you did, where would these people have come from?

Of course, there's always that slow boat from China.

~

From this scanty evidence, one could conclude that there were very few people—maybe none—south of the ice sheets prior to about 15,000 years ago: The archaeological record is extremely thin for this time and, indeed, as we will see, even if people boated down the Pacific Coast to reach Oregon and Chile 14,500 years ago, scientific evidence of substantial occupation remains scant until the Clovis show up.

Nearly 50,000 years after Australia is colonized, North America remains virtually empty of humans.

Why? This fascinating puzzle will be with us a long time. Ice and the frigid climate are considerations. And zoology still lurks in the early American thicket as a compelling argument against settlement.

CHAPTER 5

Mingled Fates of *Homo Sapiens* and *Ursus Arctos Horribilis*

Grizzly Bears as Proxy for Early Human Occupation of the Americas

AT THE SLOW DAWN OF human consciousness, fearsome beasts prowled the shadows of our primordial fires. They stalked us as dinner on the grasslands, lingered near the mouths of our Pleistocene caves and traveled with us across Beringia into the New World. They were our talismans, our monsters and our mentors. They became an embryonic chunk of our early religion and we appeased the beasts by painting their images on the innermost walls of the first human art galleries. They were lions, crocodiles and tigers. But mostly, they were bears.

Of all animals that challenged *Homo sapiens*, bears are the most humanlike. The rear-paw track looks like our own; they stand upright, have dexterous forepaws and binocular vision. Grizzlies snore in their sleep and mothers cuff their kids when they mess up. A skinned bear carries an eerie resemblance to a human corpse. Bears are called "same-sized predators." Because bears and men shared a similar omnivore diet, they lived in the same places. Both Neanderthals and *Homo sapiens* occupied the European rock shelters of cave bears, as well as brown bear, and sometimes the bones of men and bears were intentionally buried together, mingled in the same graves. Everywhere in the Northern Hemisphere where early humans traveled they found the tracks of brown bears (*Ursus arctos*). Pleistocene people followed these tracks up into northeastern Siberia and, when the human first crossed over into North America, they walked in the paw prints of bears.

The last two large omnivores to venture over from Asia to America were brown bear and human beings. The bear got here first, maybe around 60,000 years ago. People showed up later, probably sometime between 30,000 and 15,000 years ago.

Omnivores are adaptable, flexible creatures. They have to be to make a living amid changing habitats, climates and the great flux of the Ice Age. As any bear biologist knows, grizzlies are extremely intelligent, not in the same way as *Homo sapiens* with his language-inspired cognitive skills but pretty damned smart. Smarter sometimes.

Here we look at the colonization of the New World by grizzly bears in order to shed light on human prehistory. It's a useful and overlooked approach (and of course this is my own field of experience). The paleontological record of bears is important because it informs the huge time gaps in the archaeology of early America. Whereas skeletal remains of only two humans have been found in America that date older than 12,000 years ago, over thirty sets of grizzly bones have contributed both DNA and radiocarbon dates that sometimes runs beyond the limits of accurate carbon-14 dating (older than about 45,000 years old).

The study of grizzlies can contribute to the larger discussion. Since the American brown bear and early Americans occupied the same habitats and ate the same foods, they may have traveled the same colonizing routes from Beringia southward to the ice-free lands. This assessment pops up a number of times in the professional literature, though most archaeologists are uncomfortable using grizzly bear ecology to further anthropological theory. Paleontologists, too, have mixed attitudes. A brown bear scientist tracking grizzly lineages along the West Coast thought that, if a coastal route for the modern lineage of grizzlies in the lower 48 could be documented for Late Pleistocene times, it would "add considerable support for a similar (Northwest Coast) route of human colonization into the Americas." This turned out to not be the case as the lineage he was tracking never got south of the ice. On the other hand, another paleontologist concluded: "Furthermore, the first appearance of brown bears south of the LGM (Last Glacial Maximum) ice margin should not be used to date the earliest availability of an ice-free corridor for human expansion in the New World." This modesty, as we shall see, is unnecessary.

Understandably, the scientists are guarded on the usefulness of these comparisons. Having tracked, lived with and studied these formidable creatures for more than four decades, however, I firmly believe it's a big mistake to dismiss these parallel omnivore worlds. The fundamental fact, which cannot be overstated, is biotic habitats that sustain brown bears can also support humans. The only imaginable exception might be an open habitat where humans were vulnerable to short-faced bears and other giant predators but in which grizzlies could have coexisted with the megafauna.

Likewise, any route a bear could travel, early Americans could follow. The reverse is not so: Grizzlies can't pioneer the barren grounds of recently deglaciated regions as humans could have done with their bird hunting skills. Brown bears extend their collective territories by the individual home ranges of breeding females who require edible plant communities and the associated insects and small mammals that constitute tundra habitats. There are other differences: Bears evolved from ancestral carnivores but developed an elongated gut that allows them to digest both meat and some kinds of vegetation, mostly green plants in pre-flowering stages. The brown bear is not as efficient a hunter as the two-legged predator and must sleep out the winter when the only easy food is the indigestible cellulose of December's yellow grasses.

Inland grizzlies are largely vegetarian. One way to learn edible plants is to get on the ground and watch the bears; grizzlies lead you out onto the meadow or into the forest and show you what plants and berries to eat, what corms and tubers to dig. The role of the bear-teacher in the ethnographic literature is huge—widespread and circumpolar, from Norway to Siberia, British Columbia to Newfoundland. Universally, the bear is the master of seasonality, showing the people where food is found throughout most of the year.

The grizzly-mentor would have instructed all humans coming from Siberian Beringia. Much of the diet was shared. Bears eat the same berries, roots, nuts and meat that people do. On the Northwest Coast we find grizzlies eating salmon, digging clams, scavenging beached sea-mammals, chewing oysters off rocks and raiding puffin rookeries, gobbling up eggs as well as the occasional fledgling. Human coastal mariners could live off the sea-coast by eating what the bears did. Of course, grizzlies didn't occupy the entire Northwest Coast: Bears lived in the northern archipelagos and Alaskan refugia and also south of the ice in Washington and Oregon. Two different lineages of grizzlies suggest they didn't connect along the great ice fields of central British Columbia. Inland, on the tundra, in the taiga or pine forest, subsistence would have been harder. Grizzlies are usually vegetarians who supplement their diet with bugs and other animals, especially rodents including ground squirrels and voles. The first Americans no doubt watched this foraging with acute interest.

In Beringia, brown bear would have hunted winter-weakened caribou, camels or horses in spring and preyed upon all their calves come summer. Many now-extinct Pleistocene predators would have competed with grizzlies for carrion and carcasses—possibly left by human hunters, certainly by big cats, the gigantic short-faced bears and a heavy-jawed canine recently pegged as the Eastern Beringian gray wolf. All those creatures died out within two thousand years of each other at the end of the Ice Age, except for the grizzly, a more flexible and adaptable omnivore (like humans) who could survive on plants when necessary. Grizzlies on the tundra today specialize on caribou and ground squirrels to supplement a foundation vegetarian diet: horsetail, sedge, *Hedysarum* roots and Arctic cotton grass (*Eriophorum* sp.) in spring and a variety of berries in fall.

The grizzly bear as teacher would have instructed early people throughout all time periods in western North America. We have largely envisioned the earliest Americans as big-game hunters, ice-age men in skins. But this is a guess and an oversimplification. Humans have always foraged, though evidence of such activity is seldom preserved. What would be the quickest way to learn a new food group? Watch the great bear harvest the acorn crops of the live oak forests, dig up the cone caches of squirrels for pine nuts or, down in the Sierra Madres of Mexico, see how the bear uproots giant agaves just before the "century plant" shoots up its inflorescence, when the heart (*piña*) is full of natural sugars and carbohydrates. Of course humans have to cook it before they eat it. Who knows when this first happened? It also gave us pulque and, much later, mescals and tequila. The grizzly, however, had been eating agave hearts for millennia.

Where people and grizzlies both occupied the landscape, how did they get along? There is no early record. Brown bear lived in Siberia but the grizzly is more aggressive than his Asian cousin, perhaps suggesting that the Siberian side of Beringia was a less formidable place to live than Alaska.

Historically, grizzlies ranged all over western North America, eastward to the Mississippi and southward into the Sierra Madres. Since only a single Late Pleistocene specimen has been located south of the ice, we can only guess how widely they roamed: My guess is that the range of the brown bear toward the end of the Pleistocene and during Clovis times was approximately the same as its historical range with the exception of glaciated regions. The reason: All of western North America was prime habitat for grizzlies 13,000 to 26,000 years ago and there was plenty of time to expand their ursine ranges southward toward the tropics and eastward to the big river. Except for the big Pleistocene carnivores, it was pretty good ground for people too. The population of Late Pleistocene grizzlies was likely smaller than that of contact time due to competition with the short-faced bear and other scavengers.

In North America, the grizzly lived where humans did, consuming the same foods; the relationship is well documented in California where prehistoric people armed themselves with bear-lore and spears that led to mutual respect and coexistence. In the Plains and Rockies, brown bear preferred river bottoms where the native tribes also pitched their tepees. To competent and knowledgeable humans, during either prehistoric or present times, the grizzly is no threat whatsoever. It is, on occasion, a decent predator, running down and killing elk, sheep, caribou, younger bison, moose and mountain goat. Mother bears with young can be a danger to any careless human. You need to be alert, as early hunters prior to firearms no doubt were.

Grizzlies present a spectrum of social behaviors, from solitary, wary interior animals on the tundra or up the mountain who flee from humans or more dominant bears at 300 yards (a collective average), to bears on a salmon stream where a mother might leave her cubs sitting five feet from you while she goes and gets a fish. The abundance of food breaks down the mutual antagonism and intolerance of bears. This kind of tolerance, or antagonism, is extended to humans who need to be aware of sudden seasonal behavioral shifts in grizzlies—witness the fate of my friend Timothy Treadwell who lingered a week too long on a rapidly dwindling run of Alaska salmon. When food is scarce, such as Yellowstone summers when a bison dies in the rut during early August, a grizzly may defend that carcass against all invaders. A number of people have been severally mauled, even killed, under such circumstances.

The human attitude for coexisting with grizzlies was the same then as it is today—one of humility. Walk the woods and mountains like an ancient hunter: Scan the horizon, smell the air, listen to the birds. Beware of thickets and other possible day-bed areas for bears, especially mothers sleeping with their cubs. Avoid carcasses (you can smell them a mile away). Never surprise a bear. If you find yourself in a situation where the grizzly knows you're there, never run and never, ever try to climb a tree. Stand your ground and don't look directly at the bear. If you are charged (I have been a couple dozen times, almost always by sows with cubs), don't shout, move or blink. Literally. In the most unlikely event the mother bear slams into you, play dead, protect your head and neck and never fight back. We're not top dog out there. We never were.

～

A thorn of both paleontology and archaeology is that you only get what you find. You don't get the first or last individual of a fossil species or of a certain kind of archaeological site. You don't find even a fraction of the hidden data. More likely you just get slices of information representing a tiny piece of the material culture people left behind or of the skeletal remains of Pleistocene bears.

Grizzly skeletons are a good example: You almost never find the bones of a dead grizzly in a natural setting away from salmon streams or garbage dumps. Apart from those taken by modern hunters and so-called wildlife managers, it's exceedingly rare to find grizzly skeletons in the wild, compared with bones of other large animals. It is my experience—not to be confused with the scientific method—that of all the large mammals that range the wilder West, brown bear skeletal remains are the least likely to be found. That includes carnivores and omnivores such as wolverine, mountain lion, wolf and black bear. You cross several hundreds of elk and bison skeletons for every grizzly bone you find. Having roamed the wildest corners of grizzly country for most of five decades, with an eye out for such things, I've stumbled upon but four substantial fragments of bear skull, a few long-bones and the disarticulated skeleton of a young grizzly. I compare my specimens to skulls or anatomical pictures and then return the bones to wherever I found them. No doubt, I've missed a great deal. Nonetheless, I keep a tally of all the skeletal remains I've found and compare the numbers: Grizzly bear bones are the least commonly encountered of all living larger North American mammals.

Once in Yellowstone Park, during the summer after the great fires of 1988, I led my young children through the charred lodgepole forest in search of morel mushrooms. The fires had cleared out the thin understory of whortleberry, exposing a scorched soil of shattered obsidian flakes. Delicious mushrooms grew everywhere, especially along the charcoal outlines of burnt logs, like a crime scene chalking of a hit-and-run victim on the concrete. My five-year-old son, who was closest to the ground, pointed to an animal bone with a big morel that seemed to grow out of it: Half buried in the ground lay the biggest grizzly skull I had seen in the lower 48. The thin bones covering the nasal passage were burnt, one canine tooth cracked and, out through the zygomatic arch, a morel mushroom sprouted. I found another Yellowstone skull near a hot spring in which I used to warm myself during the late fall of 1968. The bear skull was in the general vicinity of an old garbage dump.

The reason grizzly remains are seldom found is probably because many of them die natural deaths in their dens. Brown bear are healthy mammals who heal well and rarely succumb to disease. They may live beyond 30 years. Human-caused mortality accounts for most modern bear deaths. Uncommonly, grizzlies are known to use caves to sleep out the winter, but suitable caves are rare and only locally available, usually in limestone topography. Before the short-faced bear succumbed to extinction 13,000 years ago, some of those caves were no doubt already occupied. So the brown bear digs dens.

For grizzlies, the approaching winter brings on hyperphagia, an urgent and often aggressive need to consume great amounts of calories in preparation for the long sleep of winter. Grizzlies dig dens, usually high on a north-facing slope of a mountain, under the roots of a sturdy spruce or pine tree, clawing out a four to ten-foot tunnel with an opening not much bigger than they are. There are variations but overall the idea constitutes an ecology of den behavior: Find a place where the snow reliably insulates the den and lingers until it's time to wake up in early spring. Modern grizzlies usually den in very remote mountain locales. Whether this apparent secrecy (they may dig their dens about six weeks ahead of the big sleep and then beeline to them during the first significant blizzards that cover their tracks) is related to humans hunting them is unknown. But if the older or wounded bear doesn't wake up, you have a very remote gravesite.

On rare occasion, brown bears have stayed out of the den all winter by appropriating the deer kills of cougars or elk kills of wolves—much like the gigantic short-faced bear must have scavenged his way through the Beringian darkness. But the grizzly is not an effective predator. This carnivore cannot subdue enough large animals to keep itself alive in winter.

My suspicion is that most grizzly deaths in the wild take place during hibernation, a natural burial in the remote country where brown bear normally dig their dens. Back when both grizzlies and short-faced bears roamed the country, brown bear were not at the top of the food pyramid. The smaller grizzlies would be at a disadvantage trying to compete at a mammoth carcass and could not survive without denning. In coldest Beringia, this competitive drawback could suffice to drive grizzlies south.

~

The practical importance to paleontology of not finding more bear fossils is that there are several gaps or hiatus in the *Ursus arctos* fossil record, big ones in the 13,000-15,000 year range: One gap is in eastern Beringia (Alaska), another along the south coast of Alaska and the other down in the lower 48 where no grizzly fossils have been found between about 13,000 to 28,000 years of age. Some scientists think the gap in the grizzly record of eastern Alaska (approximately 21,000 to 35,000 years ago) represents a genuine local extinction while the 13,000-year gap in the fossil record of Glacier and Yellowstone Park's bears most certainly does not. We just didn't find the fossil evidence of brown bears during this 15,000-year interlude.

More than 30 brown bear fossils have been recovered from Beringia, providing radiocarbon dates and mitochondrial-DNA analyses. Older dates range from 42,000 or 48,000 radiocarbon years ago, based on dates from permafrost-preserved specimens from Alaska. These are considered minimal dates as they lie near the outer limits of accuracy for radiocarbon dating, because almost all of the decaying isotope C-14 has been lost; some paleontologists think the American grizzly may have been around as long as 50,000 to 70,000 years ago. DNA analyses indicate that all three present-day grizzly lineages or clades (a group of organisms believed to have evolved from common ancestor, according to the proportion of measurable characteristics they have in common) lived together in eastern Beringia 36,000 years ago. Paleontologists think the three clades do not represent separate stages of colonization from the Old World. Beringian brown bears are recent arrivals, like humans. And they probably represent, like prehistoric humans, a single invasion from Siberia.

Today, the three lineages of grizzlies are geographically separated. Clade 2 is represented by female grizzlies (mitochondria genes are passed down only in the maternal line) on the ABC (Admiralty, Barnanof and Chichagof) islands off southeastern Alaska, Clade 3 across Alaska and northwestern Canada, and lastly Clade 4, in southern Canada and the lower 48—the bears of Yellowstone and Glacier Parks. Clade 4 was last recorded in eastern Beringia some 35,000 years ago.

The most important comparison of all with humans is how did Clade 4 grizzlies arrive in the contiguous states, south of the ice sheets? There are three conceivable routes, the conventional ones include the three same principal routes proposed for the arrival of early Americans: before the corridor closed, sometime before 23,000 years ago, down the West

Coast 15,000 years ago or southward by way of the ice-free corridor 13,000 years ago. The problem is that those dispersers should have been Clade 2 or 3, bears who lived near the Northwest Coast or at the northern end of the corridor.

All such discussion was based on the assumption, valid until 2004, that grizzlies were absent from southern Canada and the contiguous states prior to 13,000 years ago. However, as noted above, in 2002 a cranial fragment of a brown bear was located in a museum collection of fossils from fluvial gravels near Edmonton, Alberta. This well-preserved bone was subsequently dated, based on two accelerator radiocarbon dates on collagen, at 26,000 years old. Moreover, this grizzly was Clade 4 bear. The researchers concluded that, with brown bear Clade 4 south of the ice before the Last Glacial Maximum, modern bears there are likely descended from populations that were already in place.

Were humans present in America 26,000 years ago? If so, where were they living? If a grizzly made it down, could humans not have followed?

Much could be inferred from this single skull specimen. And though it's only one bear fossil, it's a good one: It would be very difficult to be mistaken about a 26,000-year-old grizzly bone with Clade 4 DNA. Where else in the world would you find one? It would be hard to make this stuff up.

This missing link of grizzly bone should lend caution to our interpretation of gaps in the fossil record. Paleontologists, who think that the hole in the fossil record between 35,000 and 21,000 years ago suggests that grizzlies were missing from eastern Beringia during this time, now have another gap in the grizzly fossil record from 26,000 (radiocarbon) to 13,000 years ago in the lower 48. I would agree that grizzlies would have a tough time competing with short-faced bears in the North when the most available food was carrion from predator (or human) kills. But the climate 35,000 to 21,000 years ago ranged all over the temperature dial; half the time it was relatively mild and then it turned very cold. Bison and caribou bones, which scientists cite as showing no such fossil breach, are a hell of a lot easier to find than grizzly fossils. We shouldn't be surprised when another brown bear fossil eventually shows up in the middle of one of those gaps.

On the other hand, if the paleontologists are correct about a genuine local extinction of grizzlies in eastern Beringia and, as theorized, it was the result of competition with short-faced bears, what does this say about human occupation of Beringia—assuming a few people might have been up there, before the LGM?

Short-faced bears are carnivores while grizzlies and humans are omnivores. It makes sense to consider *A. simus* more a scavenger than a predator. The larger short-faced bear could have out-competed the grizzly when it came to finding and feeding on carrion. Brown bear need rodents and green plant material from grasses and sedges, roots, tubers, and berries. As winter approaches and these foods become unavailable, *Ursus arctos* must dig a den and hibernate. If *Arctodus simus* could find meat from carcasses all winter long, the big bear didn't need to hibernate.

How could people survive in a habitat where grizzly bears could not? The climate from about 28,000 to 35,000 years ago was relatively warm in parts of the Arctic. The Yana people lived in western Beringia during this time. If grizzlies couldn't survive in eastern Beringia because short-faced bears kept them away from kills and carrion, maybe early Americans could have protected their bison, mammoth and caribou kills by building fires, using weapons to keep the big bears at bay and by quickly stripping and caching the meat elsewhere.

∼

What does the distribution of grizzly bear records imply for the three theoretical routes humans may have taken to get down to the lower 48 from Alaska?

Grizzlies colonize new habitats slowly. A useful way to think about grizzly society is to see it as an "Amazon" culture. The last grizzlies to survive in a fading bear ecosystem seem to be older females, mothers with daughters. Of course there has to be a male somewhere. The last grizzly killed in Colorado in 1979 was an old female with birth scars. Credible sighting of grizzlies in southwestern Colorado persisted into the 90s and beyond. I helped organize a citizen's search and a student program, which continued to look for grizzly sign in Colorado for over a decade. Similarly, female grizzlies dominate biological expansion into new territories.

Like musk ox and polar bears, grizzlies reproduce slowly. Female grizzly bears don't reach reproductive age until their fifth spring. Two cubs are the average. The mother bear keeps her cubs with her through the second year then generally weans them as two-year-olds in order for her to breed again in early June. Some females don't reproduce until they are seven or eight; some never do. Female grizzlies have been known to breed and have cubs during their early 30s, but many grizzlies, especially in heavily managed areas like Yellowstone Park, seldom survive their tenth birthday.

Grizzly bears expand their range into new habitat one female home range at a time. Sows with cubs have small ranges compared to males: twenty to thirty square miles while a boar might range over a hundred square miles or more. The quality of the habitat can shrink or inflate these numbers, less home range necessary along a productive salmon stream as compared to grizzlies on the tundra requiring greater foraging ranges. Typically a daughter sets up her own range adjacent to her mother's. A rough measure of brown bear colonization of suitable habitat unoccupied by grizzlies would be in the range of twenty to forty miles every five years or so.

Brown bear were in Beringia by about 60,000 years ago and around 30,000 years ago they came down an ice-free route to the lower states, probably via an ice-free region as opposed to a corridor between two ice sheets. That means that a pre-LGM route was also available to humans for millennia. It was probably open for many thousands of years before the glaciers advancing slammed the route shut around 20,000 or 22,000 years ago. It was an easy route with all kinds of plants and animals. Grizzlies can't live in more marginal habitats. So people had every opportunity to come south before the LGM. We know they could make it down because that grizzly did. Whether humans lived up there in Beringia during that time is of course the other question.

The presence of Pleistocene grizzly fossils on Prince of Wales Island and other southern Alaska refugia indicate an ice-free region from which humans could have launched a coastal march or float southward, once the glaciers had begun to recede some 14,500 years ago. The genetic lineage of those Alaskan grizzlies is unlike the bears of Washington and Oregon, so the implication is that big glaciers of coastal British Columbia precluded the idea of an easy walk for bears or people down that entire Pacific coast. Ancient human mariners, however, could have boated around the towering glacial heads.

Grizzlies didn't use the most recent ice-free corridor that the Clovis people are believed to have come down (though the bears live there now). When the corridor first opened, it was a bleak landscape of moraines and melt-water lakes without much vegetation. There wasn't enough food to support a bear. The earliest date for the opening of the IFC is worthy of scrutiny. The elk antler foreshaft from the Montana Clovis burial is 13,040 years old. That particular elk had to have come down the corridor: How long would it take for that passage to revegetated to the extent that elk could migrate through? It would have taken many decades or a few centuries, a lot less than the time required for a grizzly to make the same trip. Humans could have passed down the corridor even earlier, as they could have subsisted on pemmican, accompanying dogs and water-fowl; they didn't require a revegetated IFC. The corridor could have been available for human passage by 13,500 years ago.

At some point along the trail of the First Americans, with ice-age humans following the trails of brown bears and perhaps occupying Beringia in the time span of 30,000 to 14,000 years ago, the stage is set for a journey. The next question is how and when did they get south and what is the supporting evidence? This chapter glances at the possibility of people using an inland route to come down from Beringia to the lower U.S. before the LGM, prior to about 20,000 years ago. But next, with the global warming of 14,700 years ago, another very feasible route emerges from the coastal fog.

CHAPTER 6

Braving the Northwest Coast During the Time of Icebergs

Maritime Learning and Innovation in North America

DID THE FIRST AMERICANS USE the Northwest Pacific coast to accomplish the initial settlement of the Americas? What is the archeological evidence for this coastal route? Did early Americans have to figure out from square one how to live and travel along coastal environments during the last global warming of the Late Pleistocene? Certainly, their ancient ancestors in Africa, Asia and Australia knew how to build boats and exploit a maritime economy. Yet prominent American archeologists, proponents of the Pacific Coast route theory, debate if these colonizers had boats or practiced a maritime economy. Is the implication that, by 14,700 years ago, they had forgotten their ocean lessons and technology as people moved inland and into the Arctic (like Odysseus who walks inland carrying an oar over his shoulder until someone asks what the hell is that thing for)? Does the utilitarian knowledge and technology of maritime environments continue to live on in the oral histories of people for thousands of years or does this cognitive history evaporate with time lived in other habitats?

These rarefied questions about landscape learning might appear intimidating, or perhaps a bit esoteric, to an interested outsider. This discussion, however, becomes less overwhelming when I recall my own unremarkable first adventures along the ocean coasts of North America.

≈

I grew up far from the sea, about as much mid-continent as anyone can get in North America. In fact, my first glimpse of an ocean was the Arctic Ocean at Point Barrow, Alaska. Peering into the fog rolling off the ice pack, I could make out an expanse of a featureless, slate-gray ocean. Just down the beach, a beluga whale had washed up. Three days later, I found polar bear tracks. I was 21 years old.

After that sighting, I visited the Pacific Ocean, both sides. Following a couple of tours as an Army medic in the Central Highlands of Vietnam, where on jungle patrols to nearby mountaintops I could see the distant South China Sea, I was repatriated to the relative safety of the Rocky Mountains. I spent that summer in grizzly bear country, and then retreated to Arizona to plan an expedition: To backpack and film the Desert Coast of the Sea of Cortez, also known as the Gulf of California, in Sonora, Mexico. The stretch of coast I had in mind was historic Seri Indian Territory, now only partially occupied by the natives. Forty years ago, big sections of this route were wilderness.

We intended to walk it. My companion happened to be a fellow Special Forces medic, on the same A-team as myself in Vietnam, a gifted linguist and natural historian—a real find, even within such an elite group as Green Beret medics whom I had always found to be the smartest of combat grunts.

The Seri Indians were small bands of sea-going hunters and gatherers, one of only two non-agricultural tribes in all of Mexico. At the time of our little expedition, most of the people lived near two small villages on the coast opposite Tiburon Island; the Seri Indians numbered in the low hundreds, having been reduced from about 5,000 by European disease and attempts by the Spanish and Mexican armies to exterminate them. They were fierce warriors, skilled hunters who ran down deer in the mountains. Most of all, the Seri were unparalleled ocean sailors, singing navigational songs at the bows of their small carrizo-woven balsas, song lines that guided them throughout the island-studded blue waters of the Gulf of California. The leatherback sea turtle was their sacred animal. The Seri language is unique, strange even to linguists; maybe it was paddled across the gulf from the Baja Peninsula of California.

The distance between Libertad and San Carlos in Sonora is less than 200 miles in raven-distance. Besides the two attractive villages, a small chunk of development scarred this coastline. We planned to bypass this as well as another agricultural section of coast.

We'd backpack gear, food, water as well as my cheap Super-8 camera gear. Though the film was ethnographic, I would film no live Indians. Instead, the two of us would stop in Seri El Desemboque to pay our respects to the people, have dinner and spend the night before resuming our coastal foraging. The point here is that we would have to live off the land for much of our food and most of our water. The terrain was semi-mountainous with small coves, rugged headlands and sheer cliffs impossible to navigate at high tide. The landscape is incredibly beautiful. On the clearest days, you could see fin whales blowing and the many islands of the gulf. I knew next to nothing about tides or coastal travel; my companion knew more but this kind of trip was new for both of us.

A few days into our filming, we had to solve the problem of fresh water; we couldn't carry enough to see us through a week of backpacking. The Lower Sonoran Desert here received only four or five inches of rain annually, though in colder seasons the ocean dew would soak your sleeping bag. We decided to attempt to harvest the dew by sponging it off agave leaves; we didn't get much, but it took the edge off our thirst. Next, we tried an evaporation trap—a modern idea we had heard of—whereby we spread a plastic tarp with a stone in the middle over a hole in the sand into which we poured seawater. The wind destroyed this simple notion; the most we ever got was a thin pint.

Desert flora and fauna, however, thrived here like nowhere else. Eighty-foot upside-down carrots called Boojum trees grew along a coastal strip of mountains where desert bighorn sheep browsed their leaves, which sprouted after rains like ocotillo. The immense variety of plants and cacti reflected this unique, rich life zone. Wildlife flourished along the coast, with deer grazing on seaweed at low tide while coatmundi scavenged shellfish in the inter-tidal zone. We spotted a rare jaguarundi cat in the saltbush and saw the tracks of a half-dozen species of dog and cat. These animals required fresh water. The deer and javalina trails seemingly ran everywhere, but especially up the steep canyons leading up into the coastal range of volcanic mountains. Some evenings and mornings, you could see doves flying to and from the heads of these canyons.

This is how we found drinking water: After much discussion we picked larger canyons where numerous game trails coalesced into steep walled glades and birds flew. The nature of the rock was important, as it needed to hold rainwater; the bedrock couldn't be porous or too fractured. On this desert coast, a massive conglomerate was the best.

We followed the deer trails from the ocean up to narrow canyons. The torrential monsoon rains of summer had literally drilled out huge potholes in the bedrock; the ones that still held water are known as "tinajas." The bigger tinajas held water for most of the year. One survival problem was solved.

Food was the other logistical hurdle. We carried some in our packs; rice and powdered gruel of some sort, sufficient not to starve but otherwise unattractive. I had brought no fishing gear and had mindlessly planned to eat stuff from the sea, though I was clueless as to what or how to get it.

Exploring the tide pools, sleeping in the wild perfect crescents of sandy bays, we had it made. Each night, we charted the stars around a roaring fire. Soon we got hungry. The most notable, familiar creature was the rock crab, a blue or red crustacean scurrying over the bedrock at low tide. Over a week, we learned to ambush these small, tasty critters at lowest tides, coming at them from the ocean side, pinning them in tiny tidal canyons and then impaling the rock crabs with fire-sharpened sticks. We boiled them over the fire in a pot and then spent hours, usually about three, picking out the white crabmeat until we had a cup of meat. In terms of caloric intake and output, we were about breaking even.

In the course of two weeks, we had pegged the water levels at low and high tide with sticks and discerned that, here in Sonora, we had two low and two high tides a day; one was higher than the other. Every day, the tide cycle lagged behind the day before by about 50 minutes. We had started camping on the beach about the time of the new moon and now it was approaching full. The highs got higher and the lows lower with the waxing moon. That was important food-wise because we had spotted some big rock scallops at lowest tide, shells that we recognized from our beachcombing.

Beachcombing meant exploring the archeological middens located in blowouts on the sand dunes; they were essentially shell mounds containing about eight thousand years of artifacts often mixed together, from historical Seri eggshell pottery and clay turtle figurines to early Archaic Armargosa points. We were less interested in the archeology than in what the Indians were having for dinner. The shell mounds mostly consisted of clam, oyster, gastropod shells, mussel and scallops, especially rock scallops. We found many opercula (the round plate that closes the shell of a gastropod) of a particular species of snail. We figured if the prehistoric coast dwellers harvested these shellfish, they had to be good to eat.

The archeological midden was our field guide to a coastal cuisine.

We could collect the snails, which we had identified by their opercula (the same ones prehistoric people ate), at any tidal level but, even after fifteen minutes of boiling, they were bitter. We pried off the operculum and dug out the snail meat with our pocket knifes. We experimented and, within a few days, discovered that the snails under the surface of the lowest tide level were sweet, actually delicious when dipped in a sauce of olive oil, native oregano and wild chiltepines, which we picked from plants up the sandy arroyos. We located a couple colonies of rock scallop at the lowest tides. With more trial and error at collecting—you had to quickly slide a Bowie knife between the scallop and the rock to pop it off—and cooking techniques, these shellfish beds became our commissary.

At night, we drew lines and circles in the sand by firelight. The circles represented the moon, the sun and earth; the lines were crude estimates of gravitation forces. Our beach astronomy resulted from mere curiosity about a simple but central event in our daily lives—the relationship of phases of the moon and tidal cycles. We incorrectly guessed the pull of the sun was greater than it actually is, since it's big though far away. Yet you could tell, by daily observations during the two weeks, that the moon was the big dog in determining tide cycles. We spent many hours over several nights discussing the prediction of tides, a dialogue that was so unforced and obvious it seemed empirical.

~

The point of these seaside yarns is that, in my experience, both the use of marine resources, and relating their collection to the cycles of tides and phases of the moon, can be a facile matter of natural curiosity. It's intuitive: Watch what the animals do. It helps if you are hungry.

This argument about whether early Americans travelling down the northwest coast 14,500 years ago used watercraft or ate shellfish raises an older question about landscape learning, the development of human mind from its African origins and climate change.

What is the anthropological history of human adaptability, what prods our survival instincts to initiate change and can it be measured? Coastal learning shows up early in this brief chronology.

Anthropologists studying hominoid evolution suggest that it was not so much emergence onto the African savanna that drove human evolution but rapid climate change. After 4 million years of a stable environment, 200 thousand years of wildly fluctuating climate stimulated the hominoid brain to double in size and prodded the people to invent stone tools. Human evolution was nature's experiment with diversity, they say. We are all creatures of climate change.

The fossil record indicates that our species, *Homo sapiens*, arose in Africa before 195,000 years ago. (Climatologists suggest a major, long-lasting drought in sub-Sahara 200,000 years ago as a causative agent.) The emergence of the modern human mind, sometimes called "cognitive modernity," some suggest, might have flowered on a different timetable, at a much later date.

For a long time, it was assumed that symbolism, ritual, art and elaborate tool manufacture blossomed suddenly in Europe 40,000 years ago, about the time when our species first came into competitive contact with the existing Neanderthal population, who were also capable of symbolic thought. Now, scientists say, evidence of cognitive modernity has been unearthed from contexts that date much older: Anatomically modern humans may have popped up in Africa already hard-wired for modern thought processes. Previously, scientists had suggested that 50,000 years ago a genetic mutation in the brain might have rewired humans for modern thinking. No doubt, population pressure and language drove all sorts of symbolic activity and social innovation. Climatic change and geographical movement stimulated behavioral adaptation; it's sometimes difficult to pluck single activities from the collective march towards human modernity.

Thus archeologists look for "proxies" to try to identify cognitive modernity (also called "behavioral modernity"), like using seemingly unrelated stages of production in tool manufacture, art, long-distance transport of materials, grinding tools, symbolic activity or tracking of time or seasons and then linking such moments to lunar cycles. The use of language, if you could date it, would probably underlie these activities.

So, archeologists have found signs of personal adornment, African beads 70,000 years old, and 80,000 year-old bone harpoons made to spear spawning Nile catfish, a seasonal awareness of fish migrations, called "seasonal mapping," which they believe only a modern mind would be able to come up with. A red ochre workshop has been located on the Southeastern African coast that dates back 100,000 years; here the material from oxidized iron deposits had been mixed in abalone shells with fat from mammal marrow and a pinch of charcoal. The implication is that red ocher means paint for decoration or art, which in turn repre- sents an effort to communicate.

Anthropologists think the most ancient claims of cognitive moderni- ty in *Homo sapiens* come from southern Africa, where shellfish foraging and the scheduling of shellfish hunting to phases of the moon run back 160,000 years; these are perhaps the oldest proxies of all.

If cognitive modernity has been around for 160,000 years, then it's quite possible such logical capability was available when the first anatomically complete humans emerged from their African roots. Something must happen to prod this innate curiosity, to awaken our creativity to solve problems like pioneering new habitats and exploit- ing newly available resources. Otherwise, modern humans would have raced through all prehistory like jackrabbits, inventing new tools and changing behavior at every crest of a new hill; they probably always had the necessary equipment for astounding invention buried in their gray matter, but some stressor was required to draw it out. Climate change, population pressure, paths of language and the discovery of new, unfamiliar lands remain candidates that prick the inclination for innovation. The movement of ice-age explorers into unknown, uninhabited America must have been a whopping challenge and invitation for change. The First Americans had to cope with the frigid North, survive during the Last Glacial Maximum, thrive in and travel the barren landscape of the ice-free corridor when it first opened, build boats for coastal travel, co-exist with gigantic short-faced bears or invent a brand new projectile point when confronted with giant pachyderms they wanted to eat.

The implication that there was no lag between the evolution of human anatomy and the development of the modern mind is fascinating. The difficulties, however, of seashore foraging and, likewise, the complexity of human thinking capable of linking tides to phases of the moon could be overestimated. This applies to periods 160,000 years ago in southern Africa as well as 14,500 years ago on the North American coast—I doubt the harvest of marine resources and cyclic associations of tides and the moon ever required much of a cognitive revolution.

~

How would have the knowledge of exploitation of maritime ecosystems and boats have made its way to the Bering Strait?

Beginning about 70,000 years ago, Africa endured a series of climatic downturns that brought on droughts and famine. Around the same time or just before, Mount Toba in Sumatra blew—the most powerful volcanic eruption in 2 million years. A volcanic winter ensued. All the climatic upheavals around that time may have been related to Mount Toba blowing off its top. Here, with this eruption, we have the closest human-experienced parallel to today's dramatic climate change. The temperature may have dropped 6 or 7 degrees Fahrenheit during the 6 to 10 years of Toba's volcanic winter. Today, greenhouse gases in the atmosphere and oceans, plus subsequent associated cascading events affecting climate change, could warm our planet by 6 or 7 degrees in a decade or more.

Seventy thousand years ago, *Homo sapiens* populations experienced a severe bottleneck: Some genetic evidence suggests we shrank down to the low thousands worldwide; could a fatter, but distinct, bottleneck be waiting for us modern folk by the end of the century?

Humans fled parts of Africa, crossing the Red Sea in boats some 60,000 years ago. They reached Asia, sailed south to Australia by 55,000 years ago and took their boats up the coast of southern Asia, in the direction of Beringia.

Eventually, perhaps as late as 15,000 years ago, Beringians looked at the coast of what is now Alaska, and wondered if they could get down there. Maybe they remembered their previous lives as mariners.

I should state here that there is no archeological evidence of people using the northwestern Pacific coastal route—much as there is no record for humans in America's Far North, previous to the LGM. In the case of the coast, however, the route is no less plausible.

∽

A series of fortuitous accidents blew me up on the northwest Pacific coast of Canada and Alaska. Beginning with my first Alaskan king crab, wolfed down with ketchup on the old Homer Spit in 1963—a year before the tsunami took it out—this rugged coastline gradually drew me into its wild heart. Whether stumbling about as a boy paleontologist or on magazine assignment on the British Columbian coast, all jobs added up to an excuse to visit this magic, bountiful mix of sea and forest. About 25 years ago, I joined Canadian and American conservationists working with Native communities to preserve large chunks of homelands from clearcutting and mining activities. I especially thank the Round River Conservation Studies group for this opportunity. Collectively these Canadian and American organizations have helped indigenous people conserve nearly 20-million acres of intact wilderness in Heilstuk and Tlingit territory. During the last three decades, I was lucky enough to tag along on many an expedition watching these dedicated folk accomplish what I consider to be the major modern conservation success story in

North America. These fragments from my notebooks bear witness to their victory.

Notebooks, 1990s

The canoe plows through the slate gray sea in the lee of an island in an uninhabited archipelago off the coast of British Columbia. A few miles ahead, I can see the full brunt of the Pacific Ocean blast through the narrow strait that separates this island from another farther south. The late afternoon sun squints through the same gap. I hug the shoreline and start looking for a suitable cove where I can find food, firewood and set up for the night. Tomorrow I'll check out the weather and scout out a crossing down to the next island chain. It's only a couple miles of exposure, but you have to watch the tides and weather carefully. Mornings are the best time to cross the treacherous narrows, since it must be near calm to risk these open waters and safely island-hop southward. A kayak would be better, though the canoe will do. You just need lots of time: Years to make it down all the way to the Columbia River from the Arctic, maybe more years than I have left. The toughest section would be rounding Cape Caution into Queen Charlotte Strait. You'd have to hold tight to the coastline, wait out the bad weather, camp in place during the winters and live off the land the entire trip. To paddle and walk down the coast from Alaska to the lower 48—it's a dream but it could be done. Ancient people no doubt did it many times.

The people occupying ancient Beringia 14,500 years ago would have been watching this imposing, changing coastline for a long time. They had been living in the Alaskan Arctic since their ancestors arrived from Siberia. The controversial coastal route hypothesis is based on inferences. The lack of hard evidence doesn't make the route to the lower 48 any less important or fascinating. The duration of their trip south may have happened in a geologic second but those ancient voyagers didn't experience it that way. It unfolded day-by-day, people standing on a frigid beach in Alaska looking southward, watching the gigantic fronts of glaciers, which were miles across, calve into the ocean and float away as icebergs. Nineteen-thousand years ago, when the glacial advance was at its maximum, the biggest of these glaciers would have extended far out into the sea, precluding any consideration of traveling that route. By 14,500 years ago, however, the ice had begun to retreat. When the fog lifted, the people could see beyond the rivers of ice to lands the glaciers had not covered or terrain from which the ice had retreated. These ice-free areas are sometimes called refugia and, by 14,000 years ago, much of coastal southeastern Alaska and northwestern British Columbia, including large islands, were ice-free. Grizzly bears lived in these refugia, indicating that this post-glacial habitat could also support humans. Huge valley glaciers and rivers still flowed westward out of the Cordilleran ice-field into the ocean, preventing any idea of an easy stroll down the beach.

> I draw the paddle through the water, J-stroking the canoe toward a broad crescent of sand, a small cove a couple of miles from the southern tip of the island. A pebbly creek trickles into a miniature estuary. I ease the craft into the gentle surf, avoid a line of black boulders and step out onto the beach. The tide is dropping but won't be at its lowest for a few hours, just before dark. After unloading two heavy waterproof bags, I pull the canoe to the upper beach and tie it off to the root of a snag. I plunge into the forest, which is open and mossy under towering spruce trees, providing a perfect tent site. Two days ago, I watched a bear chew oysters off a big rock. Black bear swim out to these medium-sized islands but you seldom find grizzlies here. I make a cursory check for brown bear sign and, finding none, erect my small tent.

> Summer living is easy on the coast, short nights yielding to mild days with salmon, steelhead and candlefish running up creeks and rivers, berries growing off lush bushes and mushrooms sprouting from fertile soils. But fierce storms blow across the Pacific and by autumn the weather can be snarly, windy and wet with days getting shorter, dark low clouds by winter, the claustrophobic night long and cold. I probably wouldn't want to winter here, but you certainly could by building a cedar lodge and laying in smoked fish and herring roe, hunting a few deer and seals, stashing pemmican and harvesting edible seaweed and shellfish.

> I have been coming here to the outer coast of British Columbia off and on for the last couple decades. Often traveling with a few friends or family, I use whatever spare boat is available from staff and colleagues who do conservation work up here. When available, a canoe or kayak is more attractive to me than a motorized craft. Paddling solo is not recommended but I love these rare opportunities for solitude in this most fecund of New World habitats, the child's garden of gathering that is the wild Northwest Pacific Coast.

And it was likely always this way: a very easy place to live, a paradise for hunters and collectors. Pre-contact nations, such as the Nootka, Kwakiutl, Heiltsuk, Haida and Tlingit, created civilizations from rich hunting-gathering economies, arguably the only non-agricultural people ever to accomplish such a feat. Some village sites on the coast show continuous occupation for nearly ten thousand years.

The first Americans to venture down this coast were probably more nomadic, on the move, heading southward away from the big glaciers that slipped into the ocean up the coast toward Alaska. Only huge inlets, big rivers and glaciers, that in places extended out and calved off into the ocean, blocked passage southward. And, if these people had willow-woven or skin boats, they could paddle around these icy barriers during the relatively calm days of summer. They could have literally lived off the beach, digging and harvesting shellfish with a technology composed of sticks and stones—the clams, mussels and oysters seemingly limitless. At low tide, you could find abalone, crabs and edible seaweeds. Seasonally, nesting birds provided eggs and meat. If they had boats, fishing and sea-mammal hunting enter the economy.

Of course, no one knows with certainty that they had boats or that they lived off these marine resources, though it seems logical they did. Yet, debate continues among mainstream archaeologists around these two questions—if coastal people used boats or walked down and whether they practiced a marine economy. Despite the feasibility of this attractive route, archaeologists have yet to find clear evidence of early coastal travelers. The sea would have inundated much of the evidence.

> Having completed a ten-minute forage inland for a hatful of blueberries and chanterelle mushrooms, I stand on the beach near the rows of rocks and watch the rills trickle down across the shingle in the diminishing light. The lines of basaltic boulders have been stacked by ancient people, hand-moved out of the way to allow the beaching of dugout cedar-log canoes, maybe the Haida of old or marauding Nootka or some unknown people from long before. Humans have been traveling these waters and coastlines for thousands of years.
>
> The last rays of sunlight illuminate a cloudbank to the southeast hanging over the Burke Channel, a rosy brilliance that dissipates into the approaching nightfall. On the mainland, near the junction

of the Burke and Fitzhugh Sound, lies the village of Namu, where an abandoned cannery now sits on an archaeological site that dates back to 11,000 years ago.

At Namu, researchers found human tooth crowns that dated 10,000-11,000 years old. Namu, like most known prehistoric coastal villages, is located at the mouth of a salmon river. The oldest artifact from the coast, a single basalt flake dredged up from a datable context off the Queen Charlottes, is estimated to be 12,200 years old. A lithic microblade tradition elsewhere on the Northwest Coast dates only to 11,800 years ago. And that's about it: no sites as old as Clovis anywhere on the coast.

I'd better hurry; dusk is gathering over the intertidal zone and I still have work to do. I grab a folding entrenching tool and head down to the beach. Blue mussels cling to every rock surface and little eruptions of seawater mark the retracted siphons of horse clams. I decide to go for the clams. Using the small shovel, I dig down a few inches off to the side of the siphon holes until I can see the elongated shell of the clam. I pluck out the three-inch-long shellfish and repeat the process a couple dozen times. By then the light is almost gone.

I kindle a fire between beached cedar logs, set up my cooking gear and rinse the clams. The reason I chose the embedded clams over the easier-to-harvest mussels is because of the very slight risk of paralytic shellfish disease (PSD), the toxin-producing microorganisms associated with "red tides."

The horrors of PSD here are often greatly exaggerated, especially by government authorities, but they're not unknown either. Up the mainland coast about a hundred miles, the ill-fated Vancouver Expedition of 1794 camped in a place now known as Poison Cove. The hungry sailors gathered and greedily ate blue mussels. Three men became violently ill. One died. PSD is not as common in the rough, cold seas of British Columbia, where strong tides and currents disperse the concentrations of toxins, as it is in warmer, calmer oceans. Nonetheless, you need to take some precaution before you forge ahead and eat a belly-full of Pacific mussels or clams in summertime.

The first humans to travel this coast would not have had to worry about poisonous shellfish. The water would have been too cold 14,500 years ago. Melting glaciers, especially in the northern part of the coastal route, were calving into the ocean and the sea level was rising.

Shellfish can live in such environments though floating icebergs may scour the ocean bottom clear of clams down to 20-40 feet in places. In the Canadian High Arctic, north of Parry Channel, I found no clams. But the walrus did, diving forty feet below the ice, grazing off siphons of shellfish sticking up from the bottom, sucking up vast quantities of clams in a single dive. Later, Inuit biologists shared walrus clam soup with me: They had opened the siphon-stuffed stomach of a freshly killed walrus, poured the mildly acidic contents into a pot, heated it up and served it with crackers and black pepper.

Fourteen-thousand-five-hundred years ago, the earliest people to venture down the coast could have found adequate maritime food almost everywhere. Fine-grained glacial sediments would have clogged littoral zones at first, reducing productivity. But, even then, pockets of shellfish could be found along headlands and rocks and the rivers would have rapidly cleared of glacial flour, allowing salmon and other anadromous fishes to pioneer runs.

Sea level was rising at an accelerated rate 14,000 years ago. The volume of glacial melt water dumping into the Pacific Ocean probably exceeded today's flow. Sedentary coastal dwellers, if they existed, might have watched the slow rise with some alarm. Transitory travelers down the coast would not have paid as much attention.

> After gathering driftwood and stoking the fire against the gathering dew, I place the clams aside in a pan of clean water as there's grit in their digestive system. To the west, the sun has dropped behind the heavy atmosphere of the Pacific. The canoe is well stocked with condiments.
>
> By dark, the fire is roaring and a large skillet is sizzling with olive oil, the chanterelles, a sliced onion and half a head of garlic. Diced potatoes are parboiled on the side in a backpacking kettle. I drain all but an inch of water, then add the horse clams and cover the pot. As soon as the clams open, I pluck them out and add more until finished. I blend powdered milk into the mix, add the spuds, a pinch of hot red pepper, salt, and the juice of three limes. I scrape the clams from their shell with a penknife, add them to the skillet, and heat up the whole thing until the potatoes are "al dente." The wind picks up and I eat my chowder in front of a blaze of cedar driftwood listening to the roar of rising surf breaking on the outer coast.

What do scholars say about the American Pacific coastal route? This migration theory emerged from the need to explain the 12,500 radiocarbon-14 date (about 14,500 years old) from Monte Verde, Chile, a site which is too early to have been settled by Clovis or other people coming down the corridor. Without pre-Clovis dates (there are several) from south of the ice, the necessity of arguing a coastal route loses some of its urgency. Still, coming down the coast makes common sense.

Prior to the 1990s, it had been assumed that glaciers blocked the coastal route until sometime before the dates established at Monte Verde. Since then, brown bear fossils and evidence of ancient vegetation have demonstrated that some of the islands off Alaska and British Columbia, as well as some of the mainland coast, could have supported humans by about 14,500 years ago. If grizzlies lived there so could Beringians. There is a time gap in the grizzly fossil record from Alaska's Alexander Archipelago between about 30,000 to 15,000 years ago, perhaps reflecting a colder climate.

As the ice retreated from the coast, plants would have pioneered the beaches and thin soils of decomposing moraines. Pacific surface ocean winds would drive seeds, branches and logs from more southern climes into all these fertile niches. Soon, pine forests would sprout behind the broadest bays. You'd see thickets of alder, an understory of moss, lichen and berry shrubs. Outside the grizzly fossils, however, there is no hard data to support that people could have used these habitats. If it wasn't the corridor, it had to be the coast—that's the argument. As with nearly all of the fascinating unanswered questions of the peopling of the Americas, the lack of evidence has never precluded impassioned theories, some acrimony or widely argued speculation.

Mainstream American archaeologists have stated that walking down the coast was a viable option once the outer coast was free of ice and, unlike the formidable coast of today, the Late Pleistocene coastline would not have presented such a challenge. Although the sea level was lower by a few hundred feet (an average of about 360 feet), this is still a baffling statement.

Fifteen thousand years ago, rivers of ice would have occupied the channels and inlets, but as they receded ancient travelers would have had the choice of crossing on the glacier or paddling across the fjord. As a third-rate mountaineer who served a brief stint as a climbing ranger for North Cascades National Park, I've had the opportunity to cross some small glaciers in Wyoming, Washington, Montana and a couple larger ones in Alaska. Despite the unmistakable hazards of icefalls and crevasses, early people could have carefully wound their way around or through this dangerous topography, using sharpened staffs as ice axes and perhaps lengths of woven rope. But crossing a big glacier would give pause; paddling a boat across the inlet during good weather would be a safer choice.

$\sim$

Sooner or later, coastal voyagers would reach a body of water that required boats: big rivers, fjords or wide inlets. In my own experience, this includes treacherous rivers that dump directly into the deadly rough surf of the Pacific Ocean. On the outer coast of southeastern Alaska, for instance, there are two dramatic sections of coast that are cut by narrow, short but swift rivers draining from receding glaciers. I wanted to backpack along one of these stretches of coast.

From my notebooks:

> If you carried a portable raft, I pondered, you might be able to paddle or swim across the river before its current dumped you out into the pounding waves of the open ocean where you would most surely drown. It took me over a decade to get my courage up and raise enough cash for a bush plane to drop me off on this utterly stark and uninhabited coastline. I had also invested in 60 meters of thin climbing rope and a $29 small blow-up raft (lighter for backpacking than an inner-tube).
>
> The river crossing presented itself on day three. The river was fast flowing with intermittent rapids, a mere 100 feet or so across and lined with bucket-sized boulders. The only place to swim across was a relatively flat 200-yard length a half-mile above the ocean, where the river disappeared into a white waterfall of surging death waves. Just miles upstream, towered the bluish head of the glacier.
>
> I stripped down to long underwear and tennis shoes, blew up the tiny "two-person" plastic raft and stashed my backpack. The end of the rope was securely tied to a chain of big rocks. I walked upstream carrying the raft and rope, which I had tied in a bowline around my waist (I also carried a sharp knife in case of entanglement). If the

current was too much, I figured the rope would swing me back to shore downstream. A rope's length above my pack, I eased into the icy water keeping the raft on my downstream side. I had no intention of getting in the flimsy craft—it was merely a floatation device. I kicked off the rocks and sidestroked like hell for the opposite bank. In this glacial river, you had maybe ten minutes of lucidity before hypothermia set in. I stroked vigorously, eased off and let the current carry me down below a big rock, then swam as hard as I could for the far shore. The river carried me down beyond where I'd left my backpack. I was about to run out of rope and get swung back to the bank where I had started. I gave one last desperate lunge towards the far shore, grabbed a rock, then another, and slowly pulled myself free of the icy water. I was across.

Losing no time (hypothermia was still a concern), I walked upstream with the raft as far as the rope could stretch. I again secured the rope to a very big rock. The rope now formed an acute angle with the river that it spanned. I fastened the raft to the rope with a carabiner, hooked onto both with breakaway clips and slid back into the river. This time the angle was right and I virtually slid down the line back across the river to my waiting backpack.

Now I was truly cold. I took a break, kindled a fire and warmed to the task, which was getting my pack across the river. I hauled my pack and the raft as far upstream as the rope would reach and tied it off again, this time anchored to chocks that I could dislodge once across the river by yanking the rope from upstream (the sketch in my notebook shows two downstream-angled swim-routes across and one coming back for my pack). The backpack fit awkwardly inside the raft, but I tied it in best I could and hitched the package to the line. I pushed off and worked my load along the rope, getting sucked under once with all my gear. Stepping out on the far bank of the freezing river, I stripped off my soggy clothes and shook like a shaggy dog.

Here, with modern mountaineering equipment and all my fancy twentieth-century gear, I managed to cross one very small river without drowning. The first Americans down the coast would have had many hundreds of such rivers to cross, with infants, grandparents and all their belongings. We can only imagine their ordeal. But if they traveled this route, it's hard to see how they made it without boats or practicing a maritime economy.

~

Travel down the Pacific coast from Alaska would have been impossible without some use of watercraft. The route is not just a long beach punctuated by headlands so much as it is a fractured coastline of inlets, river deltas, channels and archipelagos. Above all, it is and was a glacially sculptured coastline. Some of the fjords are miles across and slice inland a hundred miles or more.

What would the now-drowned coastline have looked like? When ocean breezes blow inland from the west into a mountainous landscape at latitudes greater than about 50 degrees—British Columbia, Norway, southern Chile or New Zealand—a fjorded coastline is created. The main difference between now (we may catch up soon) and 14,500 years ago is that more water was melting out of those then-larger glaciers. That means bigger, deeper rivers to cross and fjords of salt water many miles across once the glaciers had retreated. A lowered sea level would not have significantly altered the difficulty of walking the topography of this dissected coastline. It was either a wide glacier or a deep, frigid arm of the ocean.

For some thirty years now, I've had the privilege of traveling the Northwest Coast—mostly by sailboat, zodiac inflatable, kayak or canoe. My own preferred range begins in the San Juan Islands of Washington and slides up the Queen Charlotte Channel though the Hecate Strait to the mouth of the Taku River in modern day Alaska. Especially, I love the central coast of British Columbia. For anyone who likes to camp out and live off the land, it's one of the most attractive places on earth.

But I don't always rough it: I prepare elaborate dinners washed down by fine wine on the sailboats—lucky me—of conservationist friends who live here. Dungeness crab is a major food group. Those same sailboats have fish finders and depth sounders that I study like sacred text as we cruise the archipelagos looking for remote seamounts to fish or places to drop a shrimp pot. The fjords are flooded inlets today and I frequently note that the water depth drops down 800 to 1,400 feet as you float off the edge of one of them. I mean within a few yards; the "U" shaped trench is very steep. The bottoms of these gigantic troughs are what you might expect from glacial scouring. Like chains of glacier lakes in Glacier National Park or in any glaciated terrain, deep trenches rise into higher sills where the glacier has chewed its way into the bedrock like a giant, undulating anaconda. The sills, the shallowest place in the fjords I charted, were all 500 feet or more deep. The lower sea levels of 15,000 years ago wouldn't have made any difference because it was still a river of ice or a salty swim.

Archaeological evidence of watercraft is reduced to inferences: The Red Sea was navigated at least 60,000 years ago. The first Australians crossed open water over 55,000 years ago. On California's Channel Islands, a human skeleton dated back nearly 13,000 years—you couldn't have gotten out there without a boat of some kind—and so on.

Also, boats seem almost empirical, a step of easy logic just beyond the vision: Watch the patch of floating logs and look out at the wide river that you need to cross in order to get on down the coast. Constructing a raft or a simple craft of willow or alder covered with skins or bark—once the task and vision are in place—is no more difficult than inventing a Clovis projectile point. Early people could have used willow bark ties to bind a frame of alder then cover it with sealskins. If you didn't hunt marine mammals, you could almost always find carcasses washed up on the shore. Such simple watercraft could be hauled along on foot or discarded and re-constructed at the next big river crossing or glacier head. It seems unlikely (poor preservation of organic material) that evidence of primitive boats will enter the archaeological record.

We are left to imagine the trip down the Pacific Coast.

> Fourteen thousand years ago: A long cobble beach sweeps southward into a sandy bay separated by a barren headland from yet another fjord reaching inland to the blue head of a receding glacier. People dressed in animal skins walk down the upper beach, a line of men, women and more than a dozen children, most of them carrying packs. Just beyond the big breaking waves of the Pacific, two other groups of men paddle round sealskin-covered bullboats. They tow behind them a string of unoccupied round craft—seal hides stitched around a frame of green alder branches—tied together with rawhide and strips of willow bark. Should the sea kick up in the late afternoon, the men will paddle ashore and the people will haul all their boats and other gear down

the beach southward or else just make camp until the morning when the ocean surf usually settles down. The walking is a welcome break from the forced boat trips around the huge glacial heads and the broad inlets ripped by ocean waves and tidal bores.

Back up the beach loom the rugged headlands they had paddled their boats around, after waiting a week for the weather to clear and the open-Pacific rollers to shrink into waves they could handle with their bullboats. This is the eighth year of a journey they began a thousand miles to the north, where towering glaciers splintered off into the sea and giant icebergs loomed at the mouths of saltwater inlets.

The first two years were the most treacherous. They had to paddle their little skin-boats through a literal river of icebergs, across the fronts of giant glaciers that calved directly into the ocean. Where the ice had retreated, the land was young and raw; no forests yet grew from the glacial soils. Firewood was scarce. Occasionally, the voyagers found a few logs drifted onto the beach from a distant coast in Asia or from somewhere south of the ice.

It seemed the rivers of ice would never cease dumping icebergs in their path but, by autumn of the second year, the band reached lowland regions of coast that were ice-free. Thickets of alder and thin stands of two-needle pine grew in the deepest bays. The seas soon grew too rough for marine travel in their fragile craft. In a north-facing cove, in the lee of a cliff amid lodgepole pines, the people halted as winter closed in. Far to the south, another huge glacier spewed icebergs into the dark seas of November. The key to north coastal travel in Late Pleistocene America was patience: To wait out the darkness and vicious winter storms.

With driftwood for fire and using washed-up logs on the beach as building material for constructing a substantial winter shelter out of brush and sealskins, the people settled in for the long sleep of winter. They were warm in their sleeping robes. Their elders told stories of huge legendary animals and sang songs of sunny lands around the communal fires during the interminable winter nights. When the storms abated and low tides beckoned, women and children foraged in the rich intertidal zone of the broad moonlit beach. Of all the challenges and troubles this journey presented, finding food was not one of them.

Shellfish became the stable, reliable daily food source along the coast. Oysters, clams and mussels could be found most anywhere icebergs didn't scour the ocean bottom. And even in the northern coastal regions where glaciers dominated the landscape, clams lived below the level icebergs reached under the ocean surface and mussels and oysters clung to rocky niches and headlands just miles away from the milky bays clouded with glacial flour that precluded both shellfish and runs of salmon and other anadromous fish.

The people stuffed their baskets with common shellfish at favorable tides, of which they had become acutely aware, as movement down the coast, both on foot and in the boats, depended on exposed beaches and tidal currents. Other food was available: Abalone clung to rocks at lower tide levels and crabs and octopus lived in the tide pools. The technology required for this harvest was sticks and stones; children could do most of the work. Seaweed, both brown and red, lay there for the picking; the people lightly roasted seaweed on the fire and used it to roll up crabmeat like a Pleistocene burrito.

Down the beach at the foot of a rugged headland, the hunters heard the barking of sea lions.

The third year of their journey began several weeks before the vernal equinox. At first, movement was slow as many rivers and icy inlets blocked an easy walk down the coast. The winds of spring made for a choppy ocean that was often too rough to chance a crossing of a broad inlet or glacial head by bullboat.

By summer, the party of explorers came into a coastal area with no offshore islands or archipelagos that stretched as far south as the eye could sea. Icebergs were no longer much of a problem to the boats, except near the feet of the few glaciers that ran all the way out to the ocean (one of the glaciers was huge). But an exposed coast meant dangerous surf and the people found they could travel faster on the beach, hauling the boats with them. Somewhere in the vicinity of what is today southern Alaska, they paddled across a broad bay full of small icebergs and came to a series of headlands where clouds of seabirds nested. Migratory waterfowl and seabirds had been constant companions, and sometimes dinner, throughout the journey.

Many thousands of horned puffins nested among the rocks and cliffs. Most birds were still incubating eggs laid around the summer solstice to be hatched out five weeks later. The people raided the rookery robbing the faithful bird couples of their single egg. They could simply pop the egg, in all stages of development, into their mouths or fry them on hot rocks around the fire. Three years later, they will find grizzly bears far to the south pillaging other such puffin colonies the people coveted. That winter, the voyagers found a dry cave high on a cliff. Firewood was now abundant. By January, they ran short of food and moved back to the beach and built a winter lodge, closer to the shellfish and seal colonies. Land mammals had yet to appear on the scene.

The hunters had an array of weaponry. Bolos and slings could be assembled from wave-polished beach cobbles. Spear points were fashioned from bone, sometimes notched as serrated harpoons; the hunters had found that seal and sea lion hunts were more often successful with barbed and detachable projectile points tethered by a line to a float. They could hurl their spears from the bullboats or sneak up on the marine mammals at the base of headlands. Sea lions hauled out and slept on rocks and approachable sea otter floated on their backs among giant kelp fronds with dismembered crabs on their bellies.

The people craved the fat of marine mammals to supplement their lean diet of shellfish and seaweed. Hauling their gear and bullboats along the beach or arduously padding across dangerous bays was hard work and a metabolic drain. Crossing a wide inlet could be more treacherous than avoiding icebergs. The tidal bores of several inlets sometimes combined with ocean rollers to create near-maelstroms, converging waves that crashed into one another and that could toss a skin boat like a seed spinner. At such places, the Pleistocene mariners searched out a beached log of suitable bulk and length and attached it with cross members to the bullboat as an outrigger. The paddling was made more difficult but this inconvenience was more than compensated for by the assurance of stability in the violent currents.

At night, the elders whispered around the fire. They wondered if the ocean was rising.

The people used boats for most of the seventh year as they reached what is now the central coast of British Columbia. Here lay a vast archipelago of big islands, inlets many miles across and deep channels. For the first time since pushing off from Beringia, navigation presented a challenge; an inside passage southward was maze-like on a big scale. Which channel would prove a true passage down the coast protected from the rough outer ocean? To make the wrong choice meant going up a dead-end inlet stretching hundreds of miles to the interior. Fortunately, most of dead end inlets terminated in an active glacier, shedding icebergs that served as a kind of do-not-enter sign. To scout out a water channel around an island, and differentiate a true route from an arm of the sea, could take many days. The navigators consulted the elders of the band who believed they could tell the difference by subtle variations in the flow and timing of tides. The elders contributed their collective wisdom and advised the band on big decisions, like when to go on or stop for the winter.

As they moved southward, this coastal region of islands and straits became increasingly ice-free. Pine forests rooted in the bigger low-lying valleys. Sometimes a grove of spruce and fir grew at the mouth of rivers. Some of the rivers ran clear of glacial-scour clouding and throbbed with salmon. The people found they could build rock walls at low tide in shallow river estuaries and trap the salmon when the high tide retreated. The land was rich. The people brined and smoked much salmon and scraped delicious herring roe off shallow, submerged kelp fronds and seaweed.

Whales seasonally breached in the sounds: minke, orca, gray, fin and humpback. As the people paddled across a great bay, two huge blue whales surfaced and spouted alongside the bullboats. The biggest of these graceful swimmers was nearly a hundred feet long.

The people beached their boats at the edge of a river delta in a broad, wooded bay. In the silt they saw the tracks of caribou and dire wolves. Their elders recognized the tracks from their boyhoods spent in Beringia.

By the eighth year of the journey, the people knew that they would be sharing the new land with other animals, some of the great and fierce beasts their elders had warned them about. Now the band gathered around the tracks and spoor of a legendary beast: The oval prints measured a foot across. Huge cylindrical dung boules were scattered along the animal's trail. No member of the band had seen a mastodon but all had heard the old campfire stories of these great tusked giants.

The people had hunting in their blood, an innate curiosity that drew them to constantly wonder how to harvest the largest mammals, even though they could easily survive off marine resources. But with the prey prowled the big Late Pleistocene predators. As the voyagers walked and paddled southward along the coast and into the forested regions of southern British Columbia, they encountered some creatures they knew from north of the ice and others they didn't: horses and camels, but also sabertoothed cats and bears.

Some of the men began to modify the bone harpoon points they used to hunt seals, fashioning sharpened, sturdy eight-inch-long daggers that they hafted to detachable foreshafts. These they would launch with a heavier spear shaft.

Weeks passed before the band found a young mastodon mired in a spruce swamp. The hunters crept close and delivered two spears to the ribs of the big mammal before the mastodon broke free and ferociously drove them away, swinging his giant tusks like clubs. The men followed the wounded animal for two days and finally finished him off. They started the butchering process. During the night, a gigantic bear approached the carcass, driving the men away. The men fled back to their fire and cowered as the short-faced bear fed on the remains of the mastodon only a few hundred yards away. During the night, the wide-eyed hunters listened as the huge bear cracked bones with his massive jaws. Come morning, they got the hell out of there.

The hunters returned to their ocean-side camp and told the elders of the encounter with the new fearsome creature. These older guardians of the people's future were thrown back on the quandary that had been haunting them throughout the long journey southward: Should the band try to settle down and live anywhere on this uninhabited coast or just keep on moving?

The coast was bountiful. Food from the sea was endless, the gathering easy. Dried and smoked salmon would support the band for an entire winter. Huge trees now dominated the lowland coast—spruce, fir and cedar. They could chop, burn and hollow out the center of a cedar log and make a big dugout canoe as well as use the same wood (cedar logs split true for a hundred feet using stone wedges) to build permanent long houses. The country crawled with game: huge herbivores but also new and formidable predators.

The temptation to stay in one place was countered by the presence of the big carnivores who threatened them in camp, when out foraging and, especially, at kill sites. Also, the farther south they reached, the more benign the winters became. The land was rich, beautiful and promised to only get more merciful.

By the next spring, the people were hugging the coastline of what would become the United States of America.

This fictional portrait of a journey down the Pacific coast invites some modern reflection. Despite the lack of archaeological evidence, it makes sense early Americans could have embarked on such a Pleistocene adventure, and wondering about human adaptation to changing environments is not entirely empty speculation. The climate was warming, although not as rapidly as it is today, the oceans rising and the Pleistocene mariners were moving south fast, escaping the iceberg clogged seas off southern Alaska, past the conifer forests of the Pacific Northwest and on down into the live-oak grasslands of California. They might have paused on the Olympic Peninsula to hunt mastodon or popped up the Columbia River to look at the great basaltic plateau and the high deserts of Oregon.

In such a scenario, the bands of coastal travelers would have experienced a half-dozen distinct habitats perhaps requiring specialized technical and social adaptations to hunt and forage on coastlines, in forests, along the great salmon rivers, in deserts or mountains valleys and the rich chaparral of California. I don't think the early adventurers would come down the coast to South America without wanting to explore inland and exploit the resources they found there. Unless something like a giant bear, lion or sabertoothed cat loomed up out of the interior mist, intimidating the people, forcing them to stay tethered to the relative safety of the coast and offshore islands.

∾

What are the scientific arguments for the coast? The Pacific Coast route for early or First Americans is normally presented as an alternative to the ice-free corridor route. This tale of brave voyagers and navigators has received a great deal of attention in the popular press as well as television documentaries. Advocates for the coastal entry argue that the interior corridor through the ice opened too late to account for either the date of 14,500 years old from Chile or the earliest Clovis sites (a contested 13,500 years ago) in the lower U.S. Why proponents for a coastal route feel a need to deny any use of the IFC for later Clovis people is beyond me. But there's a stack of edgy papers, many Canadian, denying human movement in the interior corridor until well after Clovis times.

Since evidence from the now-submerged coast has yet to surface, advocates for the coastal theory debunk the ice-free corridor (some authorities call it "the corridor that never was" and therefore everyone had to use the coastal route). The case against early people using this route is inferred from the dates and abundance, or absence, of fossils and pollen samples, which can add one or two thousand years to the magical time when people could have comfortably lived there. The argument is that the deglaciated corridor was inhospitable and uninhabitable for several thousand years until the land was re-colonized by plants and animals that allowed people to subsist there. Other scientists think the re-vegetation of the corridor was a more rapid process, but still too slow to allow for Clovis people to make it down in time. Many of these dates come from old melt-water lakes, where radiocarbon methods are reportedly less reliable due to bulk sample sizes and calcium, magnesium and iron in the water. In short, these radiocarbon dates are used broadly to imply the unlikelihood or impossibility of humans traveling through or living in such cruel habitats until well after Clovis times.

What this argument misses is that early pioneers during the Ice Age didn't need a landscape teeming with game and grass to pass down the corridor. Using dogs for hauling sledges and as pack animals or food, they could carry what they needed to get through the most barren sections and hunt waterfowl all the way down. Consumption of waterfowl—swans, cranes, geese and ducks—is ubiquitous from Siberia's Lake Baikal to the Yukon River in pre-Clovis times. Duck harvesting presumes use of nets—a reasonable assumption. The journey would only take a very few years. The elk-antler foreshafts buried with the Clovis child in Montana, at the southern end of the IFC, point to use of the corridor. It would have taken centuries for the corridor to grow enough grass to feed an elk. Thus, the IFC would have been open to human travelers, who could have made the migration in a few years, as early as 13,300 years ago or before. It's a serious mistake to underestimate the adaptive abilities of the founders of what some of us call the greatest human adventure of all time. A more credible argument comes from geology.

On the edge of Blackfoot country in northern Montana and southern Alberta, sprawls one of the most spectacular landscapes on earth. The High Plains sweep up to the roof of the Rocky Mountains and, at the juncture, vast grasslands run northward along the Rocky Mountain Trench. Here the landscape is studded with huge boulders as if dropped by giants from outer space. Glaciers transported these rocks, known as "erratics" and deposited them along the edge of the ice as moraines when the glacier shrunk. A technique known as cosmogenic chlorine-36 has been used to date these large glacial boulders (the Foothill Erratic Trains, as they call it) that marked the edges of the most recent glaciation in the southern part of the ice-free corridor. This sophisticated method has been used to measure the approximate length of time these quartzite erratics had been exposed to bombardment by cosmic rays by the sun.

The bottom line is that this study strongly suggested that the erratic train was a product of the ice sheets coalescing—a very important contribution. Here are some technical details: Many variables and assumptions are involved in this dating technique, such as the rate of cosmic ray production over millennia, snow cover, whether the boulder was suddenly exposed at the surface or bounced along on the glacier for a while. Even the software programs chosen to calculate the dates—that some researchers warn may result in age discrepancies in the 20 to 40% range—are contentious.

In 1997, the journal *Geology* published a paper establishing that the Foothills Erratic Train in southern Alberta is the byproduct of mountain glaciers moving eastward into the continental ice-field, and then turning southward. Most of the boulders date from 11,000 to 18,000 years ago, "erosion years" (as they call it). Of course, there are margins of error, uncalculated variables and other uncertainties that could add or subtract factors of 1,500 years or more for the youngest date and about a 5000-year margin of error for the oldest samples. Still, the pattern was unmistakable; the authors concluded that their analysis of the Foothills Erratic Train "argues strongly against the entry of humans into the Americas from Beringia during the climax of the last glaciation (LGM) via an ice-free corridor." So be it.

By 2004, however, one of the "Foothills Erratic Train" authors was pushing his conclusions to say that the ice-free corridor wasn't available for human travel until 13,500 years ago, too late to account for Clovis south of the ice, and that Clovis technology probably spread northward not southward. Later, others would use the same information to push the initial date of the corridor opening all the way down to 13,000 years ago, making the use of the corridor by Clovis people "moot" or "untenable." Once again, this research would be cited as: "the final nail in the coffin of the corridor as the route taken in the first peopling of the Americas."

The two problems in nailing down the time for traveling the ice-free corridor so tightly is that in this case geology comes up with dates that are just fine for glacial epochs or stages but shouldn't be squashed into hundred-year segments. And, secondly, that twenty-first century social scientists seem to think the first American pioneers attacked the icy passageway with the same determinism as Mall-of-America strollers.

The discussion of the Northwest Coast route is a good example of the problems faced by archaeologists when they feel a theory is probably correct but they can find no direct supporting evidence. The tendency of proponents of the coast route has been to lash out at competing alternatives or established hypotheses—in this case to discredit the route down the ice-free corridor. The real work to buttress the coastal route, the productive work, still waits up on those elevated, rebounded headlands and inland caves along the coast. Anthropologists are currently sifting away at the sediments and it is likely that they will, in time, come up with the goods to bolster their theory.

What are we left with from this discussion of a route for pioneers coming down America's Pacific Northwest Coast? Pretty much what we started with: Unexplained pre-Clovis dates from south of the ice-sheets.

CHAPTER 7

Pre-Clovis People

The Significance of People in the Contiguous American States before Clovis

In previous chapters, a great sweep of American prehistory has been presented without the benefit of much supporting hard data. I would like to think this is less the result of the wandering naturalist's imagination, which truncates my investigative method, than it is that the evidence just isn't there, at least not yet. The archaeology of early North America is a vast tundra of theoretical ground and I have enjoyed the speculative inquiries and arguments, including my own.

In contrast, in the next two chapters, "Pre-Clovis People" and the "Clovis", scientific material is available to argue the archaeological consequences: A tight fist of data has emerged to document a pre-Clovis presence and, of course, with Clovis, that record fills in exponentially. Although I have, as an informed outsider, questioned some of the professional claims, an archaeologist might come up with quite different conclusions. Because many of these particular claims directly involved the Wilsall, Montana (town near Anzick site) child burial, I have, on limited occasions, extended my cranky naturalist's neck.

❧

First a layman's perspective: I started scouring the pertinent professional archaeological literature more than a decade ago. I knew there were many sites in the Americas that were professed to be older than Clovis, older than, say 13,200 years ago. But most claims were contested, some discredited and others still actively argued. I didn't think it was a big deal. Then I began to read the professional literature: Here was the passion that electrified the field, that intensity that so fascinated me in the first place and drew me in to attempt to write up this story. Much of the early debate was framed in terms of "Clovis First" and "pre-Clovis." Now, that tension has eased a bit.

I prepared myself, went in and had a short talk with C. Vance Haynes (I had briefly studied anthropology at the University of Arizona after the war). I wondered, without actually asking, if there room for an educated layman's intrusion into the professional territory of early American archaeology. Beginning with the Clovis people (this was my impression), the archaeological record of the Americas provides a long, linear set of data for most of the remainder of that prehistory. Pre-Clovis archaeology is another story. Though rich compared to the nearly barren archaeological record prior to 15,000 years ago, the hard evidence from the pre-Clovis record is scattered through several millennia across a diverse geography from a mere handful of ancient sites. The pre-Clovis claims unfold entirely in the Late Pleistocene and most, but not all, of the dates fall within the global warming period that ended that epoch, starting about 14,700 years ago. Also, during those times, the great megafauna was still around.

I don't know a single professional archaeologist who doesn't accept at least one or two of the pre-Clovis dates. After all, DNA from coprolites in Oregon and a mammoth butchering site in Wisconsin add up to pretty solid stuff: There were people south of the ice before the Clovis projectile point shows up. But how many pioneers, for how long did they persist and in what places? Maybe these early people ghosted through the lower 48 states in tiny, vulnerable bands that died off or were killed off by carnivores before Clovis people arrived.

It is around the claims of the size of pre-Clovis populations that some of the biggest remaining arguments about the first American colonization revolve. Some very modest pre-Clovis sites (a handful of alleged tools) come packaged with theories of a growing population base settling into a myriad of habitats to explain the origins and the astonishing spread of Clovis technology by diffusion a couple of thousand years later.

One might expect to hear cries of hype and exaggeration from the professional community, but there is mostly silence. After contesting Pennsylvania's Meadowcroft Rock Shelter and Chile's Monte Verde sites for over a decade, recent pre-Clovis finds don't seem to generate such passionate criticism or as much comment as they used to, despite the fact that it's very hot academic territory.

A portion of this quietude is due to money: Most research institutions don't have much; a few have a lot. Modern investigations may utilize expensive laboratory tests, such as DNA analyses, CT scans and radiocarbon dates or optically stimulated luminescence dating. Replicating an archaeological analysis, as in the classic scientific experiment, might prove impossible because of costs and privileged access to materials or locations of excavations. The basic debate remains the significance of the pre-Clovis population.

Much as the Pacific Northwest coastal-route theory became necessary (despite the lack of evidence) to explain pre-Clovis sites below the ice, the size of the pre-Clovis population is argued by some to explain the relatively brief explosive event called Clovis that swept over the continent in as short a time as 250 years. How could they have moved so fast? There are two polar extremes in this argument, as well as some space in between.

The first position is that the continental United States was virtually unpopulated by humans, and Clovis hunters popped-out of the ice-free corridor, vaulted over the intervening hills and chased mammoth in pockets of optimal habitats until they ran out of hunting lands or hit the jungles of Central America. Few, if any, settlers resisted their intrusion or impeded their progress.

The other pole of the argument is that a preexisting human occupation of continental America provided for three thousand years of prior settlement, allowing plenty of time for that pre-Clovis population to grow in size, pioneer all the various environments and get ready to receive the Clovis lithic technology, which on arrival was passed on from one pre-Clovis settlement to the next, to people willing and able to adopt the new big-game hunting technology.

A recent corollary is that this hypothetical but presumably significant pre-Clovis population had sufficient time to develop the Clovis tool kit on their own, maybe down in Texas or the American Southeast.

This is juicy archaeological terrain, full of professional agendas, especially from discoveries during the past few years. But first a glance at some interesting so-called pre-Clovis sites.

The first candidate for a solid pre-Clovis site was Monte Verde, Chile, with dates of around 14,000 to 14,500 years ago, where butchered mastodon bones were found amid well-preserved organic artifacts. Chile is a long way from the Bering Strait and archaeologists wonder if there was sufficient time for people to come down the coast, even if they used boats. They probably didn't walk, although of course they could have. Mariners more ancient than 15,000 years ago could have colonized Monte Verde, though this is mere speculation. Some authorities, as with the bulk of pre-Clovis sites, question the accuracy of the radiocarbon dates. Skeptics claim the possibility of contamination: A reservoir effect, such as older carbon from salt water, on radiocarbon dates at Monte Verde, like coal deposits in Pennsylvania, is the knapweed-in-the-ditch that doesn't want to go away. Monte Verde is but a day or two walk from the sea, though seaweed is the only marine resource that has been found at the site. Most mainstream anthropologists accept these dates as a valid pre-Clovis occupation.

The Meadowcroft Rock Shelter in Pennsylvania's coal country is a carefully excavated site that has yielded pre-Clovis radiocarbon dates over the past 35 years. That these dates have been challenged does not mean this site's antiquity can be dismissed. The unfluted lancelate projectile points from Meadowcroft are significantly larger than those from other purported pre-Clovis sites in North America. The reported range of (contested) dates averages about 16,000 years old. This intriguing ancient camp lay close to the glacier's edge.

The mammoth kill or butcher sites at the Hebior and Schaefer sites in Wisconsin are well accepted and date about 14,500 years old.

In southeastern Virginia, Cactus Hill lies above the Nottoway River. The site is a stratified sand dune that includes a Clovis component. Below the Clovis layer, archaeologists found what appeared to be a hearth and wood dating to around 18,000 years ago, about the time of the LGM. Above the hearth were quartzite flakes and blades. A few pentagonal points were present and some trans-Atlantic-colonizing proponents claim the bifaces of Cactus Hill are transitional between European Solutrean and Clovis lithic traditions. Experts disagree on the stratigraphic evidence and some think the older dates reflect forest wild fires, not human-made hearths. They think the artifacts could have drifted down from younger strata within the site's sand dune. At any rate, the assemblage is composed of small flake-like blades that most think reflect a brief occupation.

Other claimed pre-Clovis sites include Saltville, Virginia and Topper, South Carolina, where similar flake tools and blades have been found. You could easily add another half-dozen pre-Clovis sites if you were an earnest advocate. Most all are small collections of non-descript stone artifacts. Most are in the eastern U.S.; none are fully accepted by the entire archaeological community, but some are bound to pass muster. The lingering impression is that, yes indeed, there was a small if transient pre-Clovis presence out there.

If that were not enough to convince the skeptic that pre-Clovis people are not a hallucination, some isolated butchering and kill evidence have recently emerged from dusty museum bins. A mastodon rib was found near the north shore of the Olympic Peninsula and was passed on to a Washington state archaeologist in the 1970s. It looked like it had a bone projectile point embedded in the rib. Thirty-some years later, the rib was re-examined with modern accelerator mass spectrometry (AMS) and cat-scans. The specimen was dated at 13,800 years old, 700 years before the Clovis era. The sharp bony tip embedded in the rib was placed under a high-resolution scan. Based upon this technique, the investigators believe the needle-tip of the projectile point shaft had been whittled down and sharpened. Bone protein and DNA were extracted from the projectile point and investigators think that the bone implement was mastodon but not from the same mastodon that was killed. Critics disagree, and say, no, it's not clear that the alleged bone point came from a different animal. The rib and the bone "point" could both be from the same animal. One Nevada professor doesn't think it's even a projectile point: "Elephants today," he said, "push each other all the time and break each other's ribs so it could be a bone splinter that the animal just rolled on." Butchering marks on the rib, the lead archaeologist claims, indicate a human kill, though artifacts were absent from the site where the fossil was found.

The Washington mastodon find is only a single point in the West but the geography is intriguing: The location lies between the two great salmon highways, the Frasier and Columbia River systems, and along the way of the presumed early Coastal route, which people may have used to escape the ice-fields of Beringia and western Canada.

Back in 1915, a leg bone of a Jeffersonian ground sloth was found in a peat bog in northern Wisconsin. Recent examination of this femur revealed 41 "incisions," which researchers from Canada believe could only have been made with stone tools. Artifacts and the rest of the skeleton of the sloth are missing because no one knows exactly where the femur was found. Dating indicated the femur was about 13,500 years old. Butchering marks are difficult to accurately confirm and are often contested.

∾

If this modest evidence of pre-Clovis indeed means human numbers remained small, what might have constrained population growth? How about the Late Pleistocene predators, the short-faced bear and lions that could have impeded or prevented pre-LGM (prior to around 19,000 years ago) colonization of the lower states? Would such beasts have presented less of a threat during pre-Clovis times (about 15,000 to 13,500 years ago)?

One difference is that pre-Clovis people might have had dogs, whose domestication dates back to at least this time. Dogs can carry packs, haul travois and also, in a pinch, serve as food. Hunting dogs make cornering big game such as mammoth easier, distracting the beast with the giant swinging tusks, while men moved in with spears. Dogs can save your life in a fight with a big predator.

A spear-wielding hunter, who would almost certainly die confronting a huge bear alone, can kill it with the assistance of a small pack of dogs snapping at the bear's flanks and biting his heels. I knew such a man. The Inuit was with me when I carried a spear of my own into polar bear country. I was otherwise unarmed. Most people would consider this stupid, but what the hell? Not all stories are instructional.

~

During the summer of 1991, I agreed to accompany a beluga whale expedition to the Canadian High Arctic, the island country west of Greenland. All this land was polar bear country and members of the expedition were understandably nervous. My job was to be the polar bear guy, to walk point in white bear country. My companions were Bart Lewis, Rick Ridgeway and Doug Tompkins. I was taking the bear job seriously. The pay wasn't much—the price of a plane ticket—but I figured I owed my friends and the bears a bloodless trip with no casualties.

After considerable research and some reflection, I decided to carry a spear, a well-made spear, actually a pike, mounted on a stout wooden shaft of suitable length (I measured a live, captive 1400-pound Kodiak bear in Utah to obtain the critical bear-chest-to-claw measurement). The only time this defensive weapon would be used is at the moment of truth—at the conclusion of a polar bear charge. The theory was that you anchor the stern of the shaft on the ground and aim the tip of the spear towards the narrow chest of the white bear who theoretically impales himself on his charge, if all goes according to plan. Though of course the odds are not in your favor. You'd probably die about 99% of the time.

The usual advice, which is law in many quarters, is to carry a big-bore firearm for bear. I disagreed. After all, I was recruited for this trip because of my expertise with wild bears and I had survived dozens of close calls with grizzlies, too many to buy into this fatuity about guns. Besides, I consider it unethical to voluntarily invade the last homeland of wild polar bears and then blow them away just because events might not unfold to our advantage.

The problem was I knew next to squat about polar bears. I'd seen a bunch at safe distances but hadn't interacted with them as I had with brown bears. I didn't tell my buddies this at first. I was resolute, if slightly delusional, and, if necessary, intended to use my spear to protect my friends. I considered myself responsible for all my companions should an encounter with a white bear grow ominous. After all, that was what I agreed to do: Walk point. The bedrock assumption, never discussed— that kept my carrying the spear from becoming something other than a campy joke—is that you had to be willing to die.

Of course the government disagreed: They assigned an Inuit hunter to accompany us with his bear gun. Once our Inuit checked out my spear, he decided I was serious and we became friends. As the midnight sun headed into the west, he'd share a belt of Canadian whiskey from his flask with me.

We ended up with a lot of time on our hands in this magic landscape of ice and whales and caribou and musk ox: The sun circled the landscape and our expedition leader, after attaining his beluga whale objective in the first couple hours, settled into his dome tent for days reading a thick book entitled *Reality*.

I learned to spot polar bears out on the ice pack; their coat has a slight yellowish hue, off-color from the ice. One day, I spotted seven at once. I impressed my Inuit friend by predicting when other bears were present based on the behavior of a mother polar bear and her two cubs. I knew this from my grizzly bear work. One time, we spotted another family coming from the far edge of the fjord, where hundreds of belugas rolled and narwhals occasionally crossed tusks. From my notebook:

> A quarter-mile to the south of my tent three immaculate white flecks are moving directly towards me across a contrasting canvas of brown and green tundra. They are bears. Through my binoculars I can see a mother polar bear and her two cubs. They will pass inland of my tent a hundred feet away near the foot of the bluff. I pick up my spear and head to a better vantage point, a moss-covered hummock of ancient bowhead whale bones, remnants of a thousand-year-old Thule Culture sod house.
>
> The white bear family ambles into a little ravine a hundred yards away, still heading my way. They move fluidly with unimaginable grace and beauty. Holding the eight-foot-long spear in my right hand, I grab a handful of lichen and moss with my left.
>
> Like Anteaus, the giant of Greek mythology, invincible while touching the earth, I have to be on the ground, holding tight to the world,

always sharing the land with wild animals who hold down the same
living skin of the Earth with the fierce weight of their paws.
   I await their passage.

I never got to try out a spear on a polar bear. But the Inuit had once. When he was thirteen, a bear wandered in from the ice to within sight of his village. The boy and his cousins unchained the dogs, then rushed out with their spears to meet the bear. While the dogs circled and nipped at the big bear, the boys thrust with their spears. My Inuit pal finally threw his spear and was given credit for the kill.

The Inuit hunt polar bear for fur and food, one of the few remaining traditional peoples who intentionally and regularly seek out big predators.

You wouldn't want to try this without some really good dogs.

~

Finally, two recently reported sites from the contingent states illustrate the difficulty of journalistic interpretation of solid pre-Clovis evidence, which often comes blended with overreaching contentions. I imagine this is a problem for some professionals as well as laymen like myself. Yet, when the last chaff is blown from the grains of truth, two more pre-Clovis claims emerge from the dust.

The normal procedure for reporting such sites is to publish a peer-reviewed paper, usually in *Science*, then orchestrate a series of public releases citing the material and what it means. Professional critics chip-in with reservations and sometimes outright rejection. Then the popular press reports all the finding and conclusions with a wide range of accuracy.

The most recent pre-Clovis site to hit the airways and headlines comes from Texas. A Texas A&M team excavated two blocks of alluvium about 250 yards downstream from the well-known Gault site of Clovis fame. One block produced results. The *Science* 2011 report, published as "The Buttermilk Creek Complex and the Origins of Clovis at the Debra L. Friedkin Site, Texas," claims an assemblage of 15,528 artifacts and dates between 13,200 and 15,500 years ago.

For the outsider, the interested layperson, this is perhaps the most remarkable pre-Clovis report of all: It's high-stakes poker with all the elements of extrapolated claims and meticulous interdisciplinary reasoning to lend credibility to those claims.

These early dates, the report argues, allow ample time for people to settle into the various habitats of North America, colonize South America all the way down to Monte Verde, Chile, invent the Clovis tool kit and propagate a base population through which the Clovis technology later explodes. The popular press received the unfiltered claims with a unanimous embrace.

If these dates and artifacts are eventually authenticated and claims verified, many of the mysteries of Clovis and pre-Clovis could be illuminated by this single site. It's a brilliant choice for an archaeological dig—very close to the Gault site, which has already produced pre-Clovis tools, but apparently across a symbolic fence line.

Very few establishment archaeologists have commented on this site. There should either have been a celebratory parade or the usual barrage of criticism. They probably have their reasons: Of the artifacts, 85% are microdebitage, tiny flakes less than a millimeter in any direction. Criteria exist to tell the fine flakes of microdebitage by pre-Clovis people from naturally produced particles from alluvial erosion of the local chert but I don't know if they apply here.

No radiocarbon dates are provided; apparently organic material is absent. The clay where the artifacts were found was dated by 49 optically stimulated luminescence (OSL) procedures, measuring when quartz (sand) particles were last exposed to sunlight. OSL is less precise than Carbon-14 dating and is normally used on older materials. The results are said to be variable; one prominent archaeologist called OSL precision "unimpressive," saying it has a greater margin for error than radiocarbon dates. Critics point out that the OSL-test results come from quartz particles around the artifacts but not from the "tools" themselves.

The Buttermilk Creek complex appears to be an important archaeological benchmark whose credibility won't be objectively evaluated for a while. The lay press is not particularly useful here: *Discovery* magazine says it's the sheer number (thousands) of well-dated tools—"thousands of small stone tools" repeats *Scientific American*—that is special about the Buttermilk site and that alone should settle the debate (pre-Clovis versus Clovis or Clovis-first). But less than one-half of a single percent (56) of the "artifacts" found at Buttermilk Creek are tools and they don't look much like the Clovis kit (the principal author concedes that "it's not Clovis in the strict definition"). The rest of the 15,528 artifacts are flakes. The proto-Clovis projectile point or the Clovis lithic progenitor, as smoking gun, remains at large.

But, taken as a whole, the Buttermilk Creek site, and its attendant claims, presents serious challenges to a spectrum of theories. Unless the entire site is dismissed, the mere geographical presence of humans in central Texas 15,000 years ago (assuming the contested dates hold up) challenges the notion of the danger of living with North American animal predators and difficulties of overland travel within the contiguous states.

The discovery of 56 "tools" spread over two-thousand years of Texas prehistory, can be argued either way: People, as the authors contend, settling into the environments of North America with plenty of time to develop the Clovis tool kit. Or, perhaps this central Texas site, if confirmed, just indicates another small and transient pre-Clovis population.

Also, why pre-Clovis in central Texas and not, say, in California? Of course, the Edwards Plateau is a very nice place to live: lots of springs and quality chert outcrops. The presence of humans in Texas 15,000 years ago—the date itself has not been broadly accepted—begs the question of how they got there? The Gault region is far from the standard coastal routes and most all the Pleistocene predators, including the short-faced bear, lived in those habitats during the claimed human occupation.

The final nail in the Clovis First coffin comes up again in the discussion of the Buttermilk Creek site, as it does *ad nauseum* in reportage of all the pre-Clovis sites, including Paisley Cave in Oregon. Reasonably, there should be a last, even of final last coffin nails.

In any case, it's hard to see the relevance: There were Clovis and there were pre-Clovis people. No one I know believes the Clovis people were the first humans to step into the lower 48. Although the popular press gobbles up the death of "Clovis First" drivel for monthly installments in otherwise credible magazines, the debate is artificial. There's little or no audible opposition and it's possible that Clovis and pre-Clovis have no significant archaeological or cultural connections. Pre-Clovis people may have hung around in small numbers in selected locations for several thousand years. Maybe these isolated settlements had little contact with one another—Meadowcroft never shook hands with Buttermilk Creek. Maybe none of them survived long enough to greet the Clovis hunters who possibly represent a totally separate migration down the IFC beginning around 13,300. Of course, some of the later pre-Clovis dates could overlap with the time of the Clovis big game hunters and that is fascinating. Meanwhile, we await further scrutiny from the professional archaeologists. It looks like we won't be getting to the bottom of these claims anytime soon.

∾

The second important date comes from Oregon and it's arguably the most solid of all pre-Clovis dates. Scientists have isolated human DNA from coprolites (fossilized or at least desiccated excrement) excavated during the late 2010s from the Paisley Caves in the south central part of the state, located near the ancient shoreline of a Pleistocene pluvial lake. Since genetic material cannot be directly dated, seeds and plant fibers from the same coprolite were used to determine the radiocarbon date. The stone tool assemblages are small, suggesting brief occupation. People probably dropped by the cave seasonally, on and off, for a couple millennia to camp and to take a dump. If the occupation lasted months and consisted of a dozen or more people, they might have maintained a separate latrine.

There are some problems with the interpretation of this site, but the presence of humans visiting the caves 14,000 years ago is not one of them. This inland site is almost 200 raven-miles from the Pacific Ocean, flying west over the Cascade Range via Crater Lake.

Where on earth did these people come from and why haven't we found evidence of more of them?

Analysis of broken DNA sequences from presumably human coprolites, or feces, indicated people living in this high-desert country 14,000 years ago. Archaeologists cautioned that the modern DNA from the excavators might contaminate the site, but this proved not to be the case. Others noted that the presumed human coprolites could be either canine or mixed with material from other animals. The cave produced few human artifacts. Additionally, the critics asserted that intact stratigraphy of sediments was lacking, that carbon isotope anomalies rendered the radiocarbon dates unreliable and leaching from ground water and rodent burrows contaminated DNA results. The archaeologists at Paisley maintained the morphology of the coprolites suggest human origin. This seemed questionable expertise in an unusual arena. All such scat—whether dropped in the cave by humans, other omnivores or pure carnivores—is gnawed by rodents and insects over millennia into similar looking clumps of fossil poop.

Despite these reservations, the lead University of Oregon archaeologist calls the coprolites "the perfect artifact." He maintains the contents—bone, hair and vegetation—of the ancient feces reflected "a human diet," including desert parsley, a plant that grows six inches under the ground, which further indicates these people were not explorers but living there, at home, and "very well-adapted to their environment." Additionally, they found remnant tiny threads in the coprolites suggesting: "Clearly, people were sewing their clothing, form-fitting clothing just like we have shirts, pants, those kind of things, perhaps moccasins."

Canine DNA—coyote, fox or wolf but most likely coyote—is also found in the coprolites. The archaeologist says this is probably because pre-Clovis people ate canines or the coyotes urinated on the human scat.

I'd suggest another possibility. As any desert-rat knows (Edward Abbey first pointed this fact out to me four decades ago), there's no real point in burying your poop in the desert.

"Douglas, the coyotes just dig it up and eat it anyway," Ed said.

Coyotes, and other canines, routinely consume the scat of other omnivores and even herbivores. And, in the case of a dry cave, the canine usually gets the last bite. In other words, coprolites in a desert cave could have easily passed through two digestive tracts before final deposition.

So the presence of human DNA from ancient hair or cells in a coprolite doesn't necessarily mean it is human feces nor that everything in the coprolite is something a person ate. You'd have to be certain the fossilized feces are entirely human, not canine or some other animal. I don't know how you'd establish that. As for desert parsley or other seeds or vegetables, bears, coyotes and other meso-predators routinely dig up the grass and seed caches of rodents. I've filmed both grizzlies and coyotes doing exactly this in Yellowstone National Park. If the animal in question ate the grass or seed cache and also some human excrement, you could not conclude the human diet included such vegetation.

The distinction is that DNA from human-produced coprolites is not the same thing as coprolites containing human DNA. Let's say a coyote ate human excrement some 14,000 years ago. Does this possibility invalidate either the identification of human DNA from a hair in the coyote's scat or the accompanying radiocarbon date from a seed in the canine coprolite? Absolutely not. Coyote don't eat ancient human poop, they're only interested in the contemporary stuff. People and coyotes were visiting the Paisley Cave at the same time. Paisley's most important claims stand. It doesn't matter if the coprolites are human or not.

To answer their critics, the original excavators returned with a fresh crew wearing sterile gloves and began to shift through the layers of wood-rat shit. They confirmed the stratigraphy and recovered DNA that they believe was shared by people in Asia. Though artifacts are rare, the investigators found three broken so-called Western Stemmed points from datable strata roughly coeval with Clovis times, but almost 2000 years younger than the dated coprolites; they speculate that Clovis people and the nomads at the Paisley Caves represent separate populations with distinctly different lithic technologies. The scientists believe this scant evidence suggests two migrations to North America—the Northwest coastal route for Paisley and the ice-free corridor for Clovis. Again, these far-reaching claims are based on but three fragmented projectile point stems.

One of the authors felt the need to gratuitously add that their "investigations constitute the final blow to the Clovis First theory." Why bray for attention when you are holding such serious science in your lap?

The Paisley site's most important contribution remains its credible date on human DNA samples. Though it makes no difference in the dating or confirmation that humans were present in Paisley Cave, one cannot assume all these are human coprolites; you'd have to somehow definitively prove it. On the contrary, it might be more prudent to consider most of the coprolites as canine scat that contains scavenged human feces. Reliance on morphology alone becomes inexact when the scat in question lies in a dry cave for 14,000 years. This is a small site with few artifacts that could indicate a place where a few nomadic humans just stopped in for short visits. Short-faced bear remains were found 45 miles north of Paisley at Fossil Lake, Oregon. The quest for that substantial pre-Clovis population remains elusive. Yet the geography and solid dates from this site mark Paisley as one of the most important finds in pre-Clovis archeology.

The tantalizing question of where these people came from lingers. Paisley constitutes the only confirmed pre-Clovis date from the far-interior West. Fourteen-thousand years ago, the great glaciers of North America had just begun to melt and most scientists believe the IFC still lay swallowed by ice. Early Americans could have boated down the Pacific Coast, landing sporadically to explore inland for food and to check for dangerous beasts, but those travelers would have likely consisted of small bands who already had dined at the great salmon rivers of the Northwest Coast. Perhaps a few of them came up the Columbia River, turned right at the Deschutes River following the steelhead, and ended up wandering the high-desert of Oregon looking for a place to camp and get away from the short-faced bears.

≈

In the end, one large question hangs over the peopling of the Americas like a mammoth in the corner of a sod-house: Given the ease of maritime travel down the Northwest Coast, the effortless living off the beaches and estuaries from Beringia to Chile and maybe even across Panama and around to the southeastern United States, given the richness of Late Pleistocene landscapes and the abundance of harvestable game roaming all over the Americas, why don't we see more signs of pre-Clovis people using this land?

Conventional answers include rising seas that inundated coastal sites or non-perishable artifacts that were left in locations in which archaeologists have not yet looked with productive methodologies. Either the evidence for a significant pre-Clovis presence is not there or we haven't quite figured out how to find it yet.

Once south of the ice, coastal travelers probably wandered inland. The pre-Clovis sites in Chile and Oregon lie, respectively, 36 miles upriver in Chile and some 220 miles inland on the Oregon high desert. Another cluster of pre-Clovis possibilities has been located in the southeastern United States, where people could have arrived via one of the coasts and moved inland. Other places south of the ice where people lived include a mammoth butchering site in Wisconsin and a rock shelter in Pennsylvania, both pretty far from any coast. Maybe these folk represent pockets of survivors, remnants of people who came down before the glaciers closed the route south 21,000 years ago. Others think these early settlers might be European in origin.

This isn't much evidence and what it means to many archaeologists is that the human population in the Americas was very small, nearly invisible in the record, until around 13,000 years ago. Monte Verde, Paisley and a few other sites have been occupied earlier but Clovis was a huge, explosive cultural event that dwarfs all earlier arrivals. There are other factors but that's the big one: A human presence so small it seems to hang on with the tiniest toehold in a land so resplendent in marine and terrestrial resources that it defies reproductive logic.

Coastal travelers prior to the appearance of Clovis could have settled almost any place they wished between Beringia and Chile, say south-central California, living off the chaparral, on salmon and an otherwise maritime economy, pumping out children in semi-sedentary villages. But, apparently, they did not. Maybe something kept them on the move.

People living in Arctic habitats need animal fat and protein to survive; you need to kill game, especially big game. Pre-Clovis adventurers traveling from the north into the lower states would have been in a hunting frame of mind. But again, there is scant evidence they came south, and very little to suggest that pre-Clovis people hunted big game, even with a landscape teeming with big wildlife.

What stopped them? We could be missing something huge here. The astute University of Oregon archaeological team was on to something. They had the good sense to go for the gold standard: human DNA in a datable context, in their case coprolites. What about looking for fossilized/desiccated poop in other dry caves or anywhere that conditions preserve such an ancient record. There must be such sites scattered about the West, or in eastern caves or in anaerobic peat deposits. They don't have to be human coprolites, just those of animals that lived with them, like coyotes. Pre-Clovis people were not living at the top of the food chain. Perhaps field workers could look for coprolites of Pleistocene lion and short-faced bear and, when they find them, test for human DNA.

CHAPTER 8

Clovis

The Great American Invention?

CLOVIS IS THE CAPSTONE OF North American archaeology. Clovis artifacts are the most sought-after by big-time arrowhead collectors; ten-grand might get you into the bidding for a nice big fluted Clovis projectile point, but it would take a quarter of a million dollars to buy the best one. Access to Clovis archaeology is institutional bread-and-butter for scoring large grants and long-time funding for research projects. Acquiring the rights to dig a Clovis site or study its contents is a potential industry for an academically anchored archaeologist.

The Clovis Culture, as many call it, is also the beginning and the end of a way of life in the Americas. As a colonizing event, Clovis is unprecedented in archaeological history. Its signature weapon, the Clovis projectile point, shows up all over North America within 200 years of its sudden appearance. What drove these people to move so far so fast? Of course, some archaeologists think that it was not the Clovis hunters themselves who raced across the continent but their technology, which spread throughout a substantial pre-Clovis population who already occupied the lower 48 states but for whom compelling evidence has yet to be found. But any way you look at the Clovis movement, it's so rapid that it defies standard models of dispersion and even logic. You think of things like Wanderlust.

The end, of course, was also astonishingly abrupt. The diagnostic lithic tools of Clovis disappear at the same moment as many of the great megafauna—somewhere around 12,900 to 12,800 years ago, pretty close to the onset of the Younger Dryas cooling period.

If we consider Clovis culture, or its lithic technology, as something that spread, then it must have come from somewhere—emanated from some point of origin. This is not the same as raindrops falling in a pool of still water, wavelets steadily moving out with time in perfect concentric circles. A principal argument is that Clovis people pursued and hunted mammoth. Another claim, supported by some evidence (Chapter 9), is that mammoth and other herds of big herbivores were moving around a lot during the changing climate of the last days of the Pleistocene, a time span that embraces Clovis culture. Mammoth were moving northward as well as southward, looking for productive grazing areas prior to the Younger Dryas. This implies that the spread of Clovis lithic technique and culture across the continent might not have been a simple, linear dispersion; Clovis could have emerged from the ice-free corridor—or the east coast—radiated out for a few generations, then returned following the big game.

What are the origins of Clovis? This is, perhaps, the Holy Grail of American archaeology. Very recent titles from scientifically vetted publications and esteemed publishing houses include: *"...the Origins of Clovis..."* (out of Texas); and *"...The Origin of American Clovis Culture..."* (from the Solutreans of Spain or France). I mention these publications to illustrate the profession's intense and continuing interest in Clovis origins and because we will deal with this debate in this chapter. They are relevant to discussion of the Montana Clovis child burial, especially the one called the "Solutrean theory."

Four or five possibilities have been proposed to explain the origins of Clovis. First, that the culture came from the Solutreans of Spain and France, across the Atlantic ice 18,000 years ago. Secondly, the presumably abundant though archaeologically dim pre-Clovis population developed it in place (a Texas-first theory). Third, Clovis technology originated in the American Southeast where the most projectile points have been found (the fluted points are stylistically more varied in the Southeast and it is assumed that this diversity means antiquity). Fourth, seal hunters who came down the Northwest Coast 14,000 years ago and moved inland might have invented it. And last, the oldest theory, the people coming down the ice-free corridor invented the iconic fluted spear point when they first ran into mammoth and found quality stone quarries where they could experiment and polish their lithic technique to make a weapon suitable for bringing down mammoth. This last possibility probably means that Clovis technology originated in the Missouri River country of Montana, because suitable rock quarries have not been located in Alberta.

Then the signature Clovis weapon itself: Having had the privilege to hold one of these projectile points, there is no mistaking its functional beauty. This deadly, balanced six-inch blade was designed to hunt animals bigger than bison. Rabbits and squirrels never come to mind; instead, you sense immediately that the Clovis world revolved around the hunting of proboscideans. And there is something else; the skill, the love with which the best of these points are crafted transcends function and passes into a higher domain. The remarkable aesthetics of the Clovis point suggest to some archaeologists that some of these artifacts might have been used as talismans or as holy objects to be exchanged in social contracts. Clovis people may have continued to produce these exquisite spearheads after the mammoth herds had disappeared at the far edges of their range. After all, it was only a 200-300 year time span.

Clovis people were highly mobile big-game hunters. Here and there they may have settled down temporarily as generalized hunters and gatherers, usually around bedrock sources of quality lithic material for their stone artifacts. They camped near kill sites. At least part of the time, Clovis hunters specialized in mammoth or mastodon. In the desert country of Sonora, Mexico, not far inland from where I began my expedition down the coast of the Sea of Cortez, they likely killed and ate a gomphothere, a four-tusked relative of the mastodon thought to have gone extinct in this part of North America 30,000 years ago, though not in South America. (Another cautionary note: The gomphothere, like the Aztlan rabbit, was believed to have gone regionally extinct long before the last glacial maximum. Yet here a single, credible specimen lived long enough to be hunted, or scavenged, by Clovis people. The youngest fossil record of a species [Chapter 9] doesn't necessarily indicate its extinction date.) Bison, horse and no doubt camel and tapir were also on the menu. So were turtles and smaller animals. Down in southeastern Arizona, Clovis people pit-barbequed a baby mammoth, a black bear and a rabbit.

Some archaeologists question why there isn't more evidence of mastodon hunting during Clovis times. There's a significant mastodon site with a few fluted points at Kimmswick, Missouri—along the Mississippi. Another mastodon was probably butchered near my old Boy Scout camp in Michigan. Except for Sonora, that's about it. Mastodons were probably solitary feeders. Maybe Clovis people didn't like the taste of mastodon (assuming meat was sufficiently abundant so they could afford to be picky). Mastodon browsed spruce trees. In survival situations, I sometimes snare and eat spruce grouse in early spring or late fall, when normal grouse-food such as berries is scarce. Even properly grilled, the bird is bitter fare to choke down, like chewing on an evergreen branch.

There are a dozen documented sites where Clovis killed or butchered mammoth, two (three counting the related gomphothere) more kill sites for mastodon. Most of the kill sites are located in the West.

The interpretation of these 14 sites is that of the classic argument of a half-full or half-empty vessel and is used to argue for or against humans killing off the American megafauna. The extremes of this argument contend: First, the Clovis diet was based on generalized hunting augmented with shallow pluvial lake food such as mollusks, fish, birds and lots of turtles. Or, secondly, these people specialized in hunting the biggest game on the continent. The latter contention is the essential argument for humans hunting the megafauna to extinction (commonly called "overkill").

Those against overkill would ask: What evidence? Twelve kill sites to explain the disappearance of an estimated 1,000,000 mammoth during Clovis times and two more to account for the death of another million mastodon? And no kill sites whatever (a possible kill site for now-extinct horses has been found in Alberta) for the other 33 genera of animals? They see overkill as a pathetic argument and an example of where absence of evidence really means evidence of absence. Besides, four of those 14 sites are located in southeastern Arizona, only separated by 20-some miles, and may represent a single band of Clovis hunters gone gonzo on a killing spree.

The other half of the glass: There is an astounding amount of data to prove Clovis hunters targeted and killed mammoth and mastodon. Why? Compare the Pleistocene archaeological record of kill sites in North America to the Old World. Humans hunted elephants in Africa over a time span hundreds of times longer than the 300 years during which Clovis people chased mammoths and mastodons across America. Only 12 ancient elephant kill sites have been located in Africa. In fact, in all of Europe, Asia and Africa only about 25 sites have been located where humans hunted mammoth, mastodon and elephants. And only two or three of these kill sites, they say, have human weapons associated with them. So, the argument goes, that's only 17 proven elephant-family kill sites in all Africa, North America, Europe and Asia, and 13 of those are here in the American West, with another in Missouri.

This is probably an argument, similar to those explaining the causes of Late Pleistocene extinctions or the significance of a pre-Clovis population, which no one can win for the moment. There are simply not enough smoking guns to convince everyone. But what if, for the sake of discussion, Clovis was indeed the fast-moving mammoth-hunting culture that chased the huge animals over the next pass, leap frogging to the next proboscidean refuge until they hit the edges of the continent? Admittedly, this tactic doesn't sound quite practical or even rational. Why not hunt bison instead of mammoth and mastodon? The tribe of elephant is dangerous, wickedly remembering insult and injury. A wounded mammoth swinging his gigantic tusks through a group of hunters could imperil the lives of an entire band. Bison lived where mammoth did. They were numerous and easier to locate. Bison can be herded into box canyons and driven over cliffs. Buffalo can be killed with much smaller weapons or projectile points, like Folsom points, than the Clovis people used on mammoth. But even when great herds of bison were present, Clovis hunters went on making big fluted points and pursuing mammoth. If the early hunters had been bison people, they could have settled down on the great rivers of the Plains. But they didn't. Clovis people apparently kept hounding mammoth until they ran out of them. And beyond: To the southeastern U.S. where the classic Clovis projectile point shows stylistic diversity or Texas where they could settle down near quality lithic outcrops and look for smaller game. Still, whenever they encountered mammoths, they likely went after them, as in southeastern Arizona.

Why did they move so fast (assuming there was not a significant pre-Clovis population to pass along the Clovis technology)?

Maybe these people slingshot out of the corridor so fast they couldn't slow down. Their psychic momentum combined with the leapfrog tactic of chasing mammoth herds drove Clovis to the far ocean shores in record time. Movement down the corridor (the evidence in this chapter from the Anzick site appears to point to the IFC route) could have set a world colonizing record for speed. Since there was little to eat in the recently deglaciated corridor besides migratory birds, progenitors of the Clovis people would not have lingered, but rather blasted down with their dogs and packs of pemmican. They could have made the some 2,000 miles in a handful of years. Or less. That's incredibly fast and perhaps the lifestyle lingered deep in their collective consciousness.

It's possible the Clovis culture was different from other pulses of American settlers looking for a home. Clovis people could have had little interest in settling down in one place or in subsisting on turtle soup. Just as some archaeologists have suggested the iconic Clovis projectile point may have served as a talisman or a sacred item to be exchanged in ceremonies, the object of that point—the mammoth—may have occupied a spiritual corner of the Clovis cosmos that compelled constant movement.

Before stumbling into the ditch of pop-psychology altogether, there are other notions of whacked-out cultures; the Aztecs come to mind: They seemingly lived on a doomsday psychic abyss whereby thousands, on some days many thousands, of humans had their hearts cut out to insure the continued rising of the sun. I look out at the snow-covered mountains from my insular 21st century comfort wondering how those ice-age people lived. We don't necessarily think alike.

∾

There's as much scholarly ink on Clovis as on any other topic of North American archaeology. Still, we know little of the kinds of shelters they constructed or what plants they might have used. After 13,000 years, little remains of perishable materials. The stone tool kit is well known and characterized by fluted points, bifaces and blades. There's little evidence of art. A few incised stones have been found, most of them of limestone and from Texas. Clovis probably constructed temporary structures of wood and hides; circular soil stains may mark postholes. They didn't use caves much. Clovis people left spectacular caches of stone artifacts on the land (private collectors have paid millions of dollars for some of these). About twenty caches of Clovis points, bifaces and blades have been located in North America. I visited one of these places with my friend Mark Aronson, the Iowan ecologist and educator. The Rummells-Maske site is located on an Iowan hilltop overlooking a lovely creek cutting a small valley lined with live oak trees. It is a beautiful place to leave a rare stash of finished, fluted Clovis points.

Rummels-Maske cache. Courtesy of Bill Whitaker.

A few Clovis caches were covered with red ochre. A couple of these "caches," like the Montana child, may in fact represent funeral offerings. Clovis people left few archaeological sites on the land, as you might expect from broad-ranging explorers. High-quality stone was necessary for tools, so bedrock quarries with nearby water are attractive sites; Clovis people settled some of these areas in New York, Virginia and Texas. A few professionals have made the case that such sedentary sites were the Clovis norm. Just as likely, they are unique, rich niches of quality rock and resources at the edges of Clovis migration.

There is much more information available on Clovis and my intent remains to just summarize a portion this material. What little I have to add comes from Montana, from the reason I decided to write this book—my involvement with the Wilsall Clovis child burial, also known as the Anzick Site.

~

The Anzick site is the most important archaeological site on the continent. At least that's what I told Helen Anzick, matriarch of the landowning family, when Larry Lahren, Mark Papworth and myself asked permission to re-excavate the site in 1999.

One hot summer day in 1961, Montana teenager Bill Bray followed a sluggish creek down to a fishing-hole at the base of a sandstone outcrop, overlooking Flathead Creek's confluence with a major tributary (the Shields River) of the Yellowstone River. He spotted where a badger had dug into a rodent burrow and on top of the freshly excavated soil he saw "knuckle bones" that were stained red. He reached down into the dirt pile and picked up a striking stone artifact, a six-inch-long shimmering blue chert blade flaked on both sides.

Seven years later, Ben Hargis and Calvin Sarver were using a front-end loader to remove talus for a drain field when they accidentally discovered more than a hundred red ochre-covered stone and antler artifacts in the same place where Bill Bray had found his badger burrow.

"If only," archaeologist Larry Lahren laments, "I had known about Bray's biface." To be fair, Larry was a mere teenager in 1961 and became hooked on archaeology only when he found his grandfather's Bull Durham sack full of arrowheads in 1966. But his point remains: Had Lahren or another professional been aware of the Clovis biface, four decades of denial, misrepresentation and confusion could have been avoided.

The burial of the one-and-a-half-year-old child is the oldest skeleton in the Americas, the only known Clovis burial, and the largest and most spectacular assemblage of Clovis artifacts ever found. The child was found within a lens of "gunpowder"-like sediment in the lower profile of a short cliff-face of sandy mudstones. It was not a rock shelter. The collection constitutes a complete Clovis tool kit for the killing and butchering of large mammals like mammoth. The grave offerings, all heavily stained with red ocher, consisted of eight projectile points, a minimum of six (probably eight) elk antler foreshafts, 86 bifaces—the largest over a foot-long—six flakes (unifaces), one end scraper and a couple pieces of broken flakes—about 110 artifacts in all.

Establishment archaeologists often refer to the basic Clovis tool kit as something portable: "the entire assemblage could be carried in an attaché-like pouch… compact [in] size and light weight." Should anyone want to heft the considerable Anzick tool kit, which I doubt has ever been weighed, it would have required a Schwarzenegger-tailored pouch. Many of the pieces show abrasive wear on the sides, suggesting they had been transported in hide packs or kept as heirloom items. Clovis people did occasionally leave caches of tools on the land, but the Anzick materials are all burial offerings.

∾

The Montana child burial offers clues that point to late ice-age utilization of the ice-free corridor (IFC), to finding the origins of Clovis technology and the use, or misuse, by professional anthropologists of the child's skeletal remains. It also directly challenges the hypothesis that Clovis culture originated in Europe.

The Anzick burial contains six to eight elk-antler foreshafts, a couple broken, one of them, accordingly to Lahren, intentionally. In occupied elk range, antlers are common, whether today or in the last days of the Ice Age (I saw several hundred pounds of dropped elk antler on an easy April hike in Yellowstone Park in 2012). The antlers are heavy and not the sort of baggage you'd want to carry down the IFC. Lahren and Rob Bonnichsen published an article on the Anzick foreshafts, which suggested, "that they were constructed to serve as (detachable or nondetachable) foreshafts for attaching fluted projectile points to lance shafts." The study suggests the foreshafts were mated to the fluted Clovis points (eight fluted points and eight foreshafts) and of the same age of manufacture.

The last round of radiocarbon dating on the Anzick site tested two foreshafts and a human rib. The two elk antler foreshafts both dated an uncanny 13,040 years (11,040 radiocarbon years, published in 2006). These identical dates are the best dates for this Clovis site; dating on the human skeleton may be unreliable due to modern human contamination (discussed below).

The date of 13,040 years ago marks the first known appearance of elk in the lower-48. Humans in central Alaska also hunted elk around 13,300 years ago. Elk only arrived in North America from Siberia at the very end of the Ice Age, after the global warming of 14,700 years ago and had to wait in Alaska for the IFC to open in order to get down to Montana. The habitat requirements of elk and their speed of migration are probably the same today as at the end of the Pleistocene. That would have meant a fully revegetated ice-free corridor; any elk habitat expert, hunters as well as biologists, might take a stab at the time required for elk to make that journey. I would guess perhaps at least a couple-hundred years.

The antler foreshafts provide evidence for the use of the ice-free corridor and when that route was available for human passage. If modern elk came down the corridor at least 13,300 years ago, humans could have made the same trip earlier because people wouldn't have required a completely recovered habitat in terms of flora and fauna. That would push back the date for earliest possible human travel down the IFC back to around 13,500 years ago, not far off from the original much-discredited "Clovis-first" story. An interesting question is why did the people who inhabited eastern Beringia 13,300 years ago wait until around 13,100 years ago to make the journey south. Pleistocene predators?

One may quibble with the exact timeline, but it is clear that the Anzick child's ancestors could and probably did (because of the elk antler) come down the ice-free corridor. The corollary to this observation, and the dates from the Foothill Erratic Trains (Chapter 6), is that anyone stating the IFC opened too late to accommodate Clovis migration southward is dead wrong.

As one moves southward through the IFC, the first quality lithic quarries appear south of the Missouri River in Montana, where the corridor dumped out. Larry Lahren, who did his Ph.D. work in archaeology at the University of Calgary, believes these are the northern-most quarries on the corridor route and that suitable lithic material was not available in Alberta. The stone material from which the Clovis burial artifacts have been flaked comes from five or six different bedrock sources, based on macroscopic examination by lithic experts, both archaeologists and local authorities. As an out-dated and old-fashioned geologist, I am skeptical of these petrological guesses. There is too much mineralogical variation in the local quarries within 75 miles of the Anzick site. Much of the material used to flake the Anzick tools likely came from quarries north of the site as close as 30 or 40 miles distant. Lahren has located three or four of these quarries and could, with some help, find the others. The material used for the artifacts includes moss-agate, phosphoria and porcellite. We need to test rock samples and prove where the bedrock sources of the Anzick artifacts are located, not just guess. Several laboratories can do this, including the Idaho Accelerator Center in Pocatello that irradiates and analyzes artifacts to accurately match them to their exact bedrock source. Gamma rays are zapped through the artifacts so their contained (and signature) trace elements can be measured. Previously, you could only do this in a reactor that turned arrowheads into nuclear waste. With the new method, they say, you set the objects aside "for a few days until they are radiation free."

The Solutrean theory has resurfaced in a 2012 book (*Across Atlantic Ice: The Origins of America's Clovis Culture,* by Dennis Stanford and Bruce Bradley) and has direct relevancy here—the Anzick site is referenced several times in that book. The Solutrean theory authors argue that the direction of movement of raw material for the dozen or so Clovis artifact-caches across the U.S. tracks with the larger colonization of the Americas. That is, if the lithic material used to manufacture the Montana Anzick artifacts came from eastern Wyoming, then the people were coming out of eastern regions, from the point of Clovis origins. In fact, the Anzick burial objects have been used by Solutrean advocates to bolster such an east-to-west movement: Two sources of exotic bedrock material are listed for the Anzick artifacts, moss-agate from far central eastern Wyoming, the authors of *Across Atlantic Ice* say, and another rock quarry from the southeastern corner of Montana, again to the east. Indeed, nearly all the Clovis caches referenced by Solutrean advocates indicate a west-northwest movement of raw materials; hence, the argument goes, people spreading Clovis technology from origins on the eastern coast of North America.

The big problem is that, in the case of the Anzick materials, these cited source locations of exotic stone are pure guesses. Moss-agate is also found in quarries just 40 miles north of the Anzick site. No one has tested any of the Anzick artifacts' rock sources with credible analytical geological techniques and no one knows with certainty where they came from.

A sister argument of the Solutrean theory is that because greater numbers and sizes of Clovis sites (based on Clovis points found on the surface) are located in the East, the authors' propose that "Solutrean/ Clovis" technology was introduced to America near Chesapeake Bay and came into full blossom in what is now the eastern United States. Clovis people then started exploring, the theory goes, and carrying their spear points up the big rivers and, eventually, into the American West. These scientists argue that these would have been small groups of hunters who left evidence of only small, scattered sites near big game kills. The upper Yellowstone and the Anzick site would have been about the last stop on this purely imaginative trip.

Does stylistic variation and diversity of Clovis projectile points equal antiquity? Or, assuming the iconic point (and apparently extremely successful weapon) was invented "suddenly" in the lower 48 states and spread like wildfire within a hundred years, could this stylistic variation simply reflect big game strategies and the kind of rock quarries hunters discovered? The opposite direction of Clovis movement, from Montana to the southeastern U.S. states, might also have taken place: A few hunting groups with few big projectile points, moving eastward and southward, the population growing with the generations until the a denser population making more diversified and greater numbers of spear points had time to adapt their technology to forested habitats in the East.

There's no solid data to support any of these conjectured movements. The most recent and reliable dates for Clovis culture across North America demonstrate no temporal or geographical gradient to show that Clovis came from the north, south, east or west. So far the evidence points to Clovis popping up all over the place at almost the same time. Of course, Clovis does have an origin; we just haven't found it yet.

The archaeologists who argue a Solutrean origin of Clovis imply these hunters originally were Europeans. Sometime after Clovis spread from the East, they say, a Siberian migration might have also colonized the Americas. This stance means a Clovis culture skeleton should reflect European genetic characteristics. If the Anzick child burial instead pointed to a Siberian ancestor, that would be a crucial nail in the Solutrean coffin. A lab in Copenhagen, Denmark has already analyzed the skeletal material from the Anzick child and it's said they have recovered some nuclear DNA as well as mitochondrial DNA. The report has not been released. If, as rumored, the DNA of the Anzick child indicated Asian ancestry, that might be bad news for those peddling the book on the Solutrean-becoming-Clovis theory.

Unless: The Anzick child burial is not Clovis at all and the skeletal remains are unrelated to the associated spectacular cache of Clovis materials. This happens to be the position of the authors of *Across Atlantic Ice*, who wrote: "It may be that they (the child's bones) were not associated with the Clovis Cache but were incidentally buried nearby and the red ochre staining the toddler's (that's what the Solutrean authors call the Anzick child) bones is purely coincidental."

Such a far-fetched coincidence would also allow the authors to conveniently ignore key archaeological evidence from the Anzick site that doesn't fit their model.

Future investigators should analyze the red ochre on the artifacts and compare it to the ochre on the skeleton. A 2001 paper by Smithsonian anthropologists used standard color hue charts to compare the red ochre on the bones to the stains on the artifacts; they were indeed similar. But this superficial comparison is hardly definitive. The red ochre on the stone artifacts and the human remains will require more sophisticated geologic analysis. Scientists could carry out this simple, important analysis at any time, at modest costs.

~

Curiously, no obsidian (volcanic glass) was used to make the Anzick burial objects. Yellowstone obsidian is perhaps the most desired quality lithic material in this part of the Northern Rockies, and was a valued trade good (I found a big core of what archaeologist Mark Papworth believed was Yellowstone obsidian in a Michigan red-ocher burial that dated nearly four thousand years old). Clovis caches from Idaho and points southward, westward and eastward contain spectacular obsidian Clovis points. Why didn't the Clovis at the Anzick site utilize this most-

Anzick burial artifact replicas. Courtesy of Stockton White.

prized of local lithic materials? Because, no doubt, they hadn't found the quarry yet. One lithic expert stated (an educated guess based on macroscopic impressions) that some of the Anzick artifact stone came from quarries in Wyoming, twice as far away as Yellowstone; this muddled picture of from where the material used to make the Anzick artifacts was derived is an abiding mystery. Montana is a place where river sedimentation in the flood-plains can bury Clovis-age materials. For example, just 15 miles down the Shield's River from the Anzick site, rancher Ben Stein found mammoth bones sticking out of a cut bank, according to Larry Lahren, under 30 feet of sediment; the dates from the buried mammoth remains are roughly coeval with Clovis. Outside the Anzick artifacts, less than a dozen Clovis fluted points have been found in the entire state. One of them was found at the Gardiner Post Office construction site, on the edge of Yellowstone Park, about 20-some miles north of Obsidian Cliffs, where the famous quarry is located. The spear point from Gardner was flaked from Yellowstone obsidian, a material absent from the Anzick Clovis artifacts.

What all this might suggest is that the Anzick people could have been among the earlier of Clovis people—the Clovis material at Anzick is older than the Gardner Post Office Clovis projectile point—and closer to the origins of known Clovis technology.

Could archaeologists hope to find the mother site of Clovis, the place where the iconic fluted projectile point was invented? I believe, as Lahren has long suspected, that they can. Many professional lithic experts state that one can deduce Clovis technology from its debitage—the flakes and cores left behind when you make a big projectile point. Here is a mystery—the origins of Clovis—that could be solved by modern archaeological excavation and lithic analysis of debris from those Missouri River quarries just north of the Anzick site. It's provable: Irradiate the artifacts and the quarry rock, identify the trace elements indicating the bedrock source of the lithic material and then excavate. I've visited two of these quarries and they are archaeologically intact—not yet looted or significantly damaged. If confirmed, this would lay to rest several core controversies of the archaeology of early Americans—and serve as the final nail pounded into the coffin of Texas-first or any Solutrean-origin theories.

≈

The story of the bones, the skeletal remains of the year and a half-old child, is wrapped in controversy and disrespect. For me, it is also a partisan issue: As hinted throughout this book, I carry little admiration for the archaeologists, and their institutions, who jetted up to Montana to scoop up a sample of the child's bones for radiocarbon or DNA testing without once pausing to attempt to contact a Native American or question the broader sentiment of The Native American Graves Protection and Repatriation Act (NAGPRA). As my objectivity wanders here, I will attempt to compartmentalize the facts surrounding the Clovis burial from my opinions.

In the last decade, big-shot archaeologists with the means and cash have crawled out from the edges of the continent to scoop up Anzick burial items and the skeletal remains, making repeated end-runs around NAGPRA, to publish their easy efforts in peer-reviewed articles. The locals, both professionals and laypersons—who have invested as much as 40 years in maintaining the integrity of the site and preventing the collection from being split up and sold to the highest bidder—have been ignored or pushed aside, despite their intimate historical knowledge of the site. The published results of these professional incursions have many facts wrong.

The most common misconceptions confuse the spatial relationships of the burial goods, the Clovis child's remains, and a human parietal bone found about 25 yards away on the surface of the talus slope. This skull bone, picked up by Calvin Sarver at the time of the discovery of the burial, was not connected in any way to the Clovis burial and not stained with red ochre. Ben Hargis kept it separate from the rest of the Anzick burial goods and, for a while, carried it around in the metal tool box bolted to the back of his pickup. The bleached bone dated about 10,000 years old and bears witness to the importance of this geographic landmark to indigenous people, which, as Mark Papworth suspected, probably constitutes a Paleoindian cemetery.

The most reliable telling of the discovery of the site comes from the people who found it. Once Ben Hargis and Calvin Sarver hit the burial with the front-end loader, they stopped and turned off the heavy machine. Joined by their wives that evening, the two discovers hand picked their way through the burial, recovering the remaining 95% of the collection from an area of about a yard-square. The artifacts, both stone and antler, were originally stacked on top of the human skeleton. When the bifaces were pulled out, they "clinked." Everything was covered with red ochre. Lahren interviewed these people shortly after the initial discovery. In 1999, Lahren, Mark Papworth and myself conducted another interview of Calvin Sarver and Faye Hargis at the burial site. This interview was recorded on audiotape and videotape. There's a shot of Calvin Sarver placing one large hand over the other to show how tightly the bifaces and points were stacked together on the child.

It's possible no ancient American human skeleton has been treated more shabbily than the Anzick child. The discoverers, not understanding the significance of their find, took the burial materials home and scrubbed them hard with brushes in the sink, trying to get all that red stuff off. The fragmented human remains have been separated and handled by dozens, maybe many dozens, of modern humans since their discovery. Cranial fragments were glued together with rubber cement. Everybody who came through carried off a few pieces of the child's skeleton. D.C. Taylor of the University of Montana investigated the site in 1968 and left with red-ocher-stained human remains and about twenty artifacts. When Taylor died, his son at the Northern Arizona University retained part of the child's skeletal remains. Lahren secured the return of these bones to Montana about 1998; he retrieved the twenty artifacts from Oregon State University (Bonnichsen) in 1997. Lahren and Bonnichsen also found the child's red ocher-stained clavicle (in Taylor's backfill) during their 1971 field season. The Anzick Clovis skeletal remains have been stored in scattered closets, drawers and cigar boxes for forty years, handled by many who came by and showed an interest. One should assume all the Anzick human bone samples are contaminated thorough handling by modern humans. Could this have corrupted the resultant radiocarbon dates? I don't know the answer to that. Perhaps future scientists will be able to come up with methods to determine accurate radiocarbon dates derived from the contaminated skeletal remains.

The range of carbon-14 dates on the Anzick skeleton stretch from 10,680 (in 1983) to 11,550 (in 1997). The same investigator analyzed both samples. The first date, about 12,680 years ago, is very late, some 200 years after the Younger Dryas cooling; the second date, about 13,550 years ago would mark the very beginning of Clovis. The most recent dates, reported in 2006, from the two Anzick elk antler foreshafts are identical and convincing—13,040 years old. The foreshafts lay on top of the skeleton. Contamination of the smaller human-bone fragments by modern people might explain the wide age discrepancy.

Larry Lahren has made a heroic effort to keep the collection together, to have all the missing artifacts and skeletal remains returned to the Anzick family. Several institutions have flown out to collect samples and report: The Smithsonian and the Center for the Study of the First Americans at Texas A&M in 2001; Arkansas State University in 2006, whose investigator reported, "Sarah Anzick, owner of the human skeletal remains and foreshafts, gave permission..." to test the two foreshafts and a rib; Texas A&M University in 2007 and again later to get a sample of the Clovis child for DNA analysis (the analysis is finished, completed in Denmark, but the promised report for some reason has not been released). These recent raids on the child's bones are what raise my hackles.

Of course, the Anzick child burial was found on private land. I wondered if NAGPRA applies to all institutions, so I checked; indeed, the act states that all institutions that receive federal funding are subject to NAGPRA law. Why didn't anyone at least try to contact the Native American community? Such a gesture would at least approach the minimal heart and the spirit of repatriation. It's not simple or easy; I wrote and called Crow leader Bill Yellowtail but never heard back (Sarah Anzick, I was told, tried to work with the Cheyenne when she was with NIH, apparently with similar results). But the spirit of the law works: Archaeologists in Alaska worked out a partnership with the Tlingit whereby a skeleton almost 10,000 year old was studied by the scientists and then returned to the tribe. Everyone understands a thirteen-thousand-year-old skeleton will not, in terms of cultural patrimony, have a direct connection to modern tribes.

The scientists who have run off with samples of the child's skeleton, often omit referencing their home institutions on academic publications, avoiding potential legalities if not their respective university's reputation. Why don't they try to deal with the repatriation issue, it seems easy enough to make the effort?

Heeding NAGPRA is not the only problem with the Clovis burial site. Lahren is concerned with how the line of moneyed institutions coming into play with the Anzick material is affecting "the ploys and politics" of the profession he so loves. His ex-partner, Rob Bonnichsen, used the Anzick site as a "marketing tool" for the Center for the Study of the First Americans, a center Bonnichsen founded that, in 2002, moved to Texas A&M. Currently, the Center for the Study of the First Americans at Texas A&M is funding, or has funded, any number of important Clovis and pre-Clovis projects, such as "Redefining the Age of Clovis," whereby it retested radiocarbon dates on material from Clovis-age sites, including Anzick, and recalibrated the Clovis period to about 13,100 to 12,800 years old—a valuable contribution. They also sent the Clovis child's bones to Denmark for DNA testing. The Center funded the AMS testing and scans of that "bone point" imbedded in a mastodon rib that was found in the 1970s on the Olympic Peninsula. The Center led excavations of pre-Clovis material from an area of the Gault Site (Buttermilk Creek) in Texas. The collective direction of all these investigations, papers and press releases is a broader, partisan argument for a sizeable pre-Clovis population in the Americas—an aggressively pursued and expensive agenda. They could be right. In time, with new excavations and analytical techniques, science will sift through these arguments.

Of the Clovis burial, the Arkansas State University author wrote in 2006: "To date, we are unaware of claims of affiliation or requests for repatriation made by any Native American group in the 37 years since the Anzick site was discovered." I heard those same tired, much repeated words in 2001 at a state legislature hearing by a Montana museum curator who wanted to keep the Anzick artifacts in his collection. But how should any Native American group even know of that Clovis find? The Anzick site remained a discredited secret until the past decade. 1968 seems a long time ago. Now, the site is an apparent archaeological gold mine. The child's story lives on; it's time he found an earthly home.

~

Despite the unsolved mysteries of its emergence and disappearance, the Clovis culture's colonization of the region south of the ice was astounding. The Clovis tool kit stands in contrast to all earlier American people. Pre-Clovis projectile points are uncommon and mostly small. Larger ones come from the Meadowcroft rock shelter in Pennsylvania, but this fascinating site seems an anomaly among other pre-Clovis locations: a microclimate of temperate weather so close to the glacier's edge. Some think triangular and teardrop-shaped points found in the upper Yukon basin in central Alaska, dating to about 13,300 years old, are possible progenitors of Clovis.

Clovis, on the other hand, had huge spear points, probably hunting dogs and an apparent fierce madness for killing big animals. I'm thinking of the Late Pleistocene predators again. A few of them were still around during the early days of Clovis: Some short-faced bears lived in the lower forty-eight around 13,000 years ago. *Arctodus simus* might have been diminishing in numbers, if the herbivores they scavenged and ate were also in decline—an entirely plausible scenario. (If short-faced bears had presented formidable opposition to Clovis people at their kill sites, might we have expected to find a fossil bear bone amid the scattered mammoth remains?) This evidence and material is contested and discussed in the next chapter. Pre-Clovis cultures might not have had a chance to flourish if Pleistocene lions, short-faced bears and sabertooths regularly checked their numbers by predation and discouraged their hunting and gathering forays.

The Clovis people, however, freely roamed the continent and defended their mammoth kills from whatever Pleistocene scavengers that were around. Apparently: There really isn't much hard evidence here. The Pleistocene megafauna and the Clovis culture march into the sunset at the same time, with the sudden cooling of the Younger Dryas hot on their heels. What caused this great extinction of great American animals?

CHAPTER 9

Endgame

Late Pleistocene Extinction and the Sudden Sunset of Clovis

As quickly as they appeared on the scene, Clovis hunters suddenly disappear from the Earth. Clovis people probably survived, at least some of them, but the iconic projectile point vanishes along with the last fossil records of mammoth and mastodon. Thirty-three other genera (the total number is variously believed to be between 33 and 37) of North American megafauna also died off, about half of them, some scientists believe, within a tiny temporal window of 500 years—between 13,200 and 12,700 years ago.

"Megafauna" generally means adult animals of a hundred pounds or more.

So Clovis hunters show up in Montana 13,100 years ago and less than three centuries later all these magnificent animals are dead. For the first time since emerging from Africa, modern humans participate in a major sudden extinction event (Australian Pleistocene extinction probably also took place on a longer, slower time-frame, climaxing 46,000 to 48,000 years ago). The degree to which people were responsible is debated among a mix of causes including climate changes, introduced diseases and an asteroid impact.

The Late Pleistocene extinctions are one of the great mysteries in human history.

Fierce debates rage around what knocked off the megafauna. To blame it all on *Homo sapiens* begs the dark question of the nature of the beast: Are we humans the murderous, homicidal brutes so deservedly kicked out of Eden, ready to duke or nuke it out to the end of the Earth or are we at heart the deeply sentient beings we sometimes suspect ourselves to be, capable of tolerance, generosity and the kind of empathy it will require to deal with our own terminal crisis—the irreversible global climate change we likely have already brought upon ourselves?

The crucible of stewing ideas to explain Late Pleistocene extinctions in the America transcends archaeology, paleontology and even science. At bedrock, it is a question of philosophy.

Montana's great novelist A.B. Guthrie wrote, "Each man kills the thing he loves. No man ever did it more thoroughly than the fur hunters." He's talking beaver. Our own mountain men, though not Indians, were about as aboriginal as Europeans would ever get.

The "Rewilding" movement of modern conservationists who want to restore some of the megafauna, such as substituting elephants for mammoth, to North America's plains, points out that our standard of ecological excellence—the state of America in the discovery year of 1492—actually represents an impoverished landscape, already plundered of its magnificent wildlife 12,800 years ago (later Native Americans also significantly altered the land with fire and agriculture). The sudden disappearance of North America's large mammals lurks as a metaphor for the way humans regard the earth, the rapaciousness with which we consume and destroy the beauty of nature.

One of the possible answers to the extinctions is a theory called "overkill," a hypothesis formulated some forty-five years ago. The overkill theory postulates a band of highly skilled hunters bursting out of the ice-free corridor before 13,000 years ago, then radiating out blitzkrieg-style in a widening wave of megafaunal slaughter until they reach the oceans and Central America. The band might have started out with as few as 100 Clovis hunters but the population reproduced rapidly. The prey—mammoth, mastodon, ground sloth, camels and horses—had not been hunted by humans before and were therefore naïve and easy to kill.

Other explanations for the sudden disappearance of the great American megafauna include climate change, diseases introduced by people or their dogs and an asteroid smashing into the Great Lakes region. And there are combinations of these factors to explain the extinctions. Maybe climate and disease started to disrupt animal interactions by fragmenting populations or altering their ecological relationships. The big critters might have already begun their decline and human hunting just added the finishing coup-de-grace.

Many Late Pleistocene extinction arguments, however, tend to favor single theory causes.

As a tapestry for discussion of megafauna extinction, what evidence is available? The Clovis hunters were on the scene from 13,100 to almost 12,800 years ago. Older dates are possible but contested. The commencement of climate warming occurred 14,700 years ago and was interrupted by the sudden chill of the Younger Dryas, the temperature dropping 27 degrees Fahrenheit in Greenland. The date for the onset of the Younger Dryas is 12,880 years ago. It lasted nearly 1,300 years, when the current Holocene epoch begins. These are relatively precise dates from ice cores. The opening of the ice-free corridor dates back to at least 13,300 or 13,500 years (as evidenced by the elk antler foreshaft from the Montana Clovis child burial) and might have eventually provided passage for icy blasts of Arctic air blowing down from Alaska thereby cooling the climate.

More elusive is the timetable of when the megafauna disappeared; the 34 to 37 genera of animals didn't all die off at the same moment. But a growing body of radiocarbon dates and stratigraphic data seems to be narrowing the time most of the big animals succumbed to a 500-year window. Others think the megafauna were on the way out 1,500 years before Clovis even showed up and the extinction event was spread over several millennia. The most vulnerable animals, then and today, tend to be large ones with low reproductive rates.

Dates for this extinction are approximate, as the fossil record doesn't necessarily reflect the last surviving members of their taxa; a few mammoths survived until six to four-thousand years ago on islands near the Bering Strait. Some radiocarbon dates suggest the extinction was staggered over time, with horses and camels dying out 200 years before mastodons and mammoths. Yet carefully excavated Clovis sites in southeastern Arizona show that all these animals, along with tapirs, dire wolves and American lions, disappeared from the Earth at the same time, about 12,900 years ago. A black, carbon-rich layer of sediment that dates to just less than 12,900 years ago marks the disappearance of Clovis from the archaeological record. In the southwest, they are called "black mats" and probably correspond to a rise in the water table following a regional drought. Clovis tools and mammoth (and other now-extinct animals) bones appear just below the layer but not within it or above it. Black-mat layers shows up at about 70% of Clovis sites across America. In the Southwest, they are probably caused by algae growth; in the East, organic carbon from increased moisture or burning is more common. All seem synchronous with the abrupt onset of the Younger Dryas (YD) cooling.

The beginning of the YD does indeed sound like a doomsday event: The sudden sunset of Clovis culture, the extinction of three-dozen genera of animals including birds and small mammals. Some authorities think North American human numbers plummeted; at a number of formerly Clovis locations, the fluted blade technology reemerges as a smaller bison-hunting weapon, wielded by a perhaps smaller population of Folsom hunters. Other archaeologists dispute population decline by counting the number of post-Clovis Paleoindian projectile points and sites, which they argue shows that the immediate post-Younger Dryas population remained stable. But at most stratified Clovis sites, the carbon-rich layer seems to be the end of human activity for a long time.

∾

When polled, archaeologists usually point to a mix of climate change and over-hunting to explain the American extinctions. It's difficult to isolate climate. After a great deal of discussion, the question—who or what caused the American megafauna extinctions at the end of the Pleistocene—might come down to a belief system.

The quandary: When humans colonize a new island, even a great big one like North America, do the island's creatures, shortly thereafter, begin to go extinct? Every time and unavoidably? Good books are written about island biogeography and most of the evidence points in the direction of the two-legged predator. But the deal is not quite sealed. The question I often ask myself is if I think the great American Pleistocene animals would have gone extinct without the effects of changing climates (I also wonder if the Clovis phenomenon could have ever taken place without our continental warming)? After all, a number of climate reversals came and went and only the last one saw a massive extinction, the difference being human hunters on the scene.

The reasoning, at least mine, is circular: the Allerød warming of 14,700 years ago perhaps unleashed the beginning of a human invasion, which climaxed with Clovis. But the warming didn't directly kill off the megafauna. The Younger Dryas cooling at 12,880 years marks both the functional end of Clovis and the megafauna, but this return to a cold climate should have been weather to which the American megafauna was already adapted. Clovis, as manifested by its lithic technology, disappeared because there were no more very large mammals left to hunt: Or because they knocked off the last mammoths? It's a snake eating its tail.

Some things don't add up. Did the short-faced bear impede settlement of North America prior to Clovis? If so, did its population decline prior to Clovis? And if they did so, what caused the bear population to plummet? No doubt because of a lack of available prey and the carcasses of herbivores, which, again, could be due to climate change, habitat fragmentation and/or hunting pressure. Any time you knock a top-predator out of an ecosystem, cascading chaotic destabilization results; we don't begin to understand how these removals might have affected the Late Pleistocene habitats. Why did the Clovis colonization succeed so expansively whereas the pre-Clovis occupation apparently just sizzled in place? Was it the Clovis tool kit, dogs and their hunting techniques or the absence of Pleistocene bears and lions that paved the way? There's plenty of residual confusion here and the best available evidence doesn't quite solve the problem.

Paleo-extinction is still a formidable mystery.

The discussion is more challenging.

$\sim$

Why hunt the biggest animals? A mammoth is a lot of meat and outweighs a puny human by a hundred times. Bringing down a great big one means prestige, a feast, a party and maybe reproductive opportunities. Pursuit of mammoth herds fits the leapfrog notion of Clovis colonization: Theory has it that the Bølling-Allerød warming period at the end of the Ice Age fragmented the remaining mammoth populations into oasis-like pockets of habitat, which made them vulnerable to hunters. Kill sites for the horse, ground sloth or camel have not been found (Clovis tools from a cache found in 2008 near Boulder, Colorado tested positive for horse and camel protein residue) but mammoth bones, as opposed to smaller mammals, are more often recognized as something exotic due to their massive size and the big fossils are probably more resistant to erosion, thus more often preserved intact.

It's argued that these kill sites demonstrate that Clovis people were for the most part the fast moving adventurous elephant hunters out-of-Alaska. In a few productive ecological niches near the end of the southern extension of non-desert habitat (where they may have run out of easily hunted big game), Clovis people found good foraging habitat and chert quarries, where they could settle down and live like generalized hunters and gatherers. In short, the fact that a few sites indicate a generalized subsistence pattern of living doesn't refute the idea that, elsewhere, highly mobile Clovis hunters were chasing diminishing herds of big game. The size of the Clovis population, its growth and genetic diversity remains elusive, despite mind-boggling models.

Some of those models imply that Clovis hunters, armed with their signature weapon, followed their preferred prey, the mammoth, from one valley to the next, perhaps leap-frogging between mammoth refugia, which might explain the explosive nature of their colonization. Detailed knowledge of local plants and landscapes were unnecessary. Regional variations of the classic Clovis point didn't appear because the entire band picked up and left in pursuit of the next herd of big mammals. High quality lithic materials from quarries were important. They camped in the open, and didn't use caves much. As game ran out and the band was subjected to stress, the theory goes, they could choose between changing territories and switching resources. Instead of taking the time to learn about available new food resources, they just moved on.

One should step back before brushing off human-overkill to explain the end of the Pleistocene extinctions. Severe climate shifts had taken place several times in the Americas without any noticeable megafauna die off. The climate was even hotter, with higher seas, during the inter-glacial period 125,000 years ago. The difference during Clovis times: *Homo sapiens* were on the scene.

After Clovis and the extinctions of the huge animals, the fluted blade tradition lingered as a smaller, finer bison-hunting tool (Folsom). Big game hunting persisted, especially in the West for a couple thousand years and then, nearly everywhere, people seem to have settled down and lived as generalized hunters and gatherers until agriculture arrived.

~

Humans, or their dogs, coming from Asia could have introduced viral diseases that might have jumped species boundaries and infected the American animal populations. The viral diseases AIDS (HIV), SARS and the "Spanish" flu of 1918 jumped from animals to humans; maybe the reverse took place. Zoonotic diseases can be bacterial, such as bubonic plague, although the scary ones today tend to be viral and of the RNA variety, which can mutate and adapt rapidly. Cross-species transmis-sion tend to occur when humans and animals are increasingly coming into contact with one another, as they might have during the climatic-induced wildlife movements of the Late Pleistocene or, ever so alarm-ingly today, with 7 billion humans beginning to flee the consequences of global warming.

One would think, however, in such a broad spectrum of Pleistocene species that went extinct, resistance to specific diseases would preclude wholesale die off. After all, the mostly viral microbes would require a high degree of transmissibility and fairly strong dose of deadly virulence. But again, for such infectious agents to cross the boundaries of so many different species in such a brief time period seems unlikely. The highest population estimates for Clovis people (generous estimates are less than a million hunter/gatherers; more cautious guesses place Clovis numbers at around 50,000), and their dogs, still paint a stark landscape where interactions between humans and animals were extremely limited—compared to our surging two-legged hordes spilling into our last 21st century wildlife habitats.

If such a plague unfolded 13,000 years ago, the evidence would be hard to find. You would need fossil DNA from extinct animals, and then attempt to identify the DNA or RNA of a pathological virus amid that of the host. Modern molecular genetic techniques can sometimes do this sort of thing but I understand you need to know what you are looking for, and we don't.

An extinction-causing "Comet Impact Theory" was introduced to the public in 2007, proposing that an extraterrestrial impact, probably a comet or asteroid, exploded in the atmosphere somewhere over Canada north of the Great Lakes 12,900 years ago. No crater has been found (the ice-sheet could have absorbed the blast), but the team believes the products (micro-meteorites, elevated iridium levels and nano-diamonds) found in black mats across the country indicate an impact event and associated widespread burning away from the ice-sheets. Others say these are ordinary components of cosmic dust steadily raining down on Earth and samples provided by the team for radiocarbon dating were found to be contaminated. Whether this theoretical impact could have been solely responsible for the abrupt cooling event or merely a regional contributor to the Younger Dryas is another question.

In 2012, the same scientists expanded their search and found melt-glass materials in Pennsylvania, South Carolina and Syria, which they believe resulted from a large cosmic body impacting the Earth's atmosphere.

Many authorities doubt the impact theory altogether: A recent study demonstrated that black (charcoal) mats simply mean global weather changes. The same markers may simply indicate a wetland environment. Wildfires occur with each episode of climate change. In the case of the onset of the Younger Dryas, no comet is necessary to explain widespread burning; the sudden cooling killed off vast pine forests, providing sufficient fuel (before the pine needles fell off) to account for lightning torched large-scale fires and the concentration of charcoal particles.

At any rate, the impact, if it indeed occurred, didn't by itself cause the extinction of the Pleistocene megafauna, which some think had already been in decline for 2,000 years. Sloth survived the theoretical impact in Caribbean islands, mammoth endured in Alaska, the South American megafauna was still going strong 12,500 years ago and grizzlies and buffalo are still here.

∾

What information could be drawn from these events? Clovis culture blossomed in a time of global warming, rising oceans and melting ice. Mid-North America was prime wildlife territory for fast-moving big game hunters. All this ended suddenly 12,880 years ago with the severe cooling of the Younger Dryas. Scientists like to point out that nearly every animal over 220 pounds died off and only animals weighing less than that survived this extinction. A notable exception was the grizzly (along with modern bison, moose, elk, caribou, musk ox, polar bear and chunky humans).

Why did some large mammals survive while the other megafauna disappeared close to the Younger Dryas cooling? Grizzlies are relatively recent arrivals to North America, having crossed over from Siberia some 50,000 to 70,000 years ago. The omnivorous bears evolved alongside humans. Elk and modern moose are even more recent immigrants; they show up in Beringia about the same time as the earliest pre-Clovis dates and thus knew humans as formidable predators. As a hunted species, elk may have gained a slight advantage as prey animals. Bison and caribou might have survived because of their vast numbers and musk ox and polar bear tends to live in cold, remote habitats.

Grizzlies are generalized omnivores, much like humans. The paleontological literature mentions a few 13,000-year-old brown bear fossils from the lower-48; if these specimens still exist, isotopic analysis on bear bone might indicate what they were eating before and after the time of the "black mat" formation giving us a hint of what daily life was like during this extreme climatic event.

∾

Mushrooms are fruiting bodies of fungi, which are neither plant nor animal. When you kick a mature mushroom or puffball, the little cloud of dust you see consists of millions of spores. The abundance of dung fungus, *Sporormeilla spp.*, spores has been proposed as a proxy for the presence or absence of the large Pleistocene herbivores—the fungus indicates megafauna. Dung fungus refers to the kind of fungus that must pass through an animal's digestive tract to complete its life cycle. The microscopic spores of fungus are then picked up by winds and deposited on lakes, where they sink to the bottom. Thus, scientists have studied sediment core samples from lakes in New York, the West and Indiana. Some twenty species of *Sporormeilla* are hosted in the feces of mammals, birds and probably other creatures. But mostly, these scientists think, in mammal shit. Recent waterholes grazed by modern cattle have been studied for comparison and, yes, the relative abundance of *Sporormeilla* does indicate when cows were present.

The results from the eastern lakes, contrasted to the western sediment cores, are not identical. In California and Colorado, the frequency of dung fungus falls off sharply at about 12,800 years ago. This would be the expected result just before or at the extinction of big mammals at the time of the Younger Dryas onset. An increase of charcoal particles, indicating ancient wildfires, follows this spore decline.

The New York lake sediment cores, however, show *Sporormeilla* declining about 14,500 years ago, some 1,400 years before Clovis people show up in the archaeological record. After the disappearance of the megafauna (as represented by the spore decline), the pollen samples indicate that the vegetation underwent change, perhaps because the megafauna wasn't around to put grazing pressure on the land, chewing and altering it into communities for which there are no modern analogs. The spore decline is followed by charcoal, more evidence of burning.

Much the same pattern is claimed for the cores taken from a kettle pond, Appleman Lake, in Indiana. The *Sporormeilla* spore frequencies are treated as an absolute proxy for the herbivore megafauna and its local abundance, it is suggested, represents North America. The first decline in Sporormeilla spores begins 14,800 years ago and continues to a final decline a thousand years later, when the spore-to-pollen ratio dropped below 2 percent, signaling a disappearance of the large herbivores from the area. These results are argued by the involved scientists as a collapse or local extinction event, unrelated to any comet/asteroid-impact theory and probably unrelated to climate change. As the authors state: "Megafaunal populations collapsed from 14,800 to 13,700 years ago, well before the final extinctions and during the Bølling-Allerød warming period. Human impacts remain plausible, but the decline predates Younger Dryas cooling and the extraterrestrial impact event proposed to have occurred 12,900 years ago."

So, what caused the decline? We are left, the scientists say, with humans coming into the lower states 1,500 years before Clovis, hunting the megafauna towards extinction and setting fire to the landscape. Of course, there is no solid archaeology to support the presence or activities of such effective hunters. Similar interpretations are made for the New York record: An initial blitzkrieg-like hunting spree 14,500 years ago, they argue, precipitating a sharp decline in the megafauna followed by a buildup of non-grazed fuel with subsequent burning.

What comes to mind is that these events, as interpreted from *Sporormeilla* spore frequencies as a proxy for the presence or absence of the megafauna, may represent very local occurrences. The steep decline of dung fungus spores might just mean the herds of megafauna simply moved away to another region or habitat. In comparison studies of present day cattle/*Sporormeilla* relationships, the amount of spores you recover depend on how far from the shore of the lake or pond you are; the highest densities of *Sporormeilla* occur next to the bank, not in the middle of the pond. The scientists' broad interpretations overrun the confined evidence. The fossil record should reflect this abrupt decline to confirm these conclusions. If you could sample a hundred or a thousand lakes across North America, some very interesting patterns might emerge.

Canadian paleontologists have found evidence that, just prior to the Holocene, mammoth and horse dominated the Late Pleistocene landscape in central Alberta. Perhaps these now-extinct mammals returned north to beat the heat.

The initial dates for estimates of *Sporormeilla* declines in New York and Indiana, 14,500 and 14,800 years ago, are very close to the beginning of the Bølling-Allerød global warming event at 14,700 years ago. Climate change could have influenced vegetation types and megafauna distribution in patterns for which we have yet to find evidence, triggering the herds to migrate and leave the area.

Climate warming, at 14,700 years ago, may have also accommodated the movement of human hunters throughout the continent, beginning a chain of environmental changes, nudged along by big game hunting during a time of weather upheavals. Thus we have the scarce and scattered pre-Clovis "hunters." It's one thing to accept that a few people made it south of the glaciers before Clovis, but quite a stretch to think there were enough of them to directly contribute to an overkill of the megafauna.

In New York, the mastodons lasted until 13,000 years ago, by direct dating on bone; they didn't go out 14,500 years ago, or 13,700 years ago, as the fungal-spore-data indicated. The huge grazers of the Late Pleistocene routinely chewed the habitat into a relatively fireproof landscape, but when they departed, became extinct or migrated, fuel accumulated. Lightning struck and grass and forest fires scorched the land. The presence of humans to set the fires is not required to explain all the charcoal in the sediment layers, although early hunters might have torched a few blazes on their own.

What the heavily extrapolated *Sporormeilla* dates from Indiana might mean, rather than representing the entire continent, is that it was a time of climatic flux in grassland habitats and the great herds were moving around a lot.

≈

One last note on extinction theories: A very recent study on the tooth wear of Pleistocene lions and sabertooth cats mired in the La Brea tar pits indicated there wasn't much change in dietary habits of these predators at the end of the Pleistocene. To the scientists, this means carnivore life was no tougher or more difficult than before humans showed up; declining number of prey was not a primary cause of extinction for these large cats. When prey resources are limited (due to human hunting or climate change), the assumption is that carnivores consume carcasses more completely, chewing up the bones and breaking or leaving distinctive wear-marks on their teeth. Another assumption of the study is that specimen percent-abundance is a proxy for social behavior—that the 80 mostly male, large cranium American lions were less social than the 2,000 sabertooth cats. Another prominent American archeologist has alternatively suggested that *Smilodon fatalis* may not have been especially bright, as the sabertooth was the second most frequently trapped carnivore (4,000 dire wolves) in the La Brea tar pits.

At bottom, this study is an argument against human overkill or climate change decreasing predator opportunities in southern California during the Late Pleistocene—because there is no evidence of increased "carcass utilization" by the big cats at La Brea. Another explanation is that, with so many dire wolves and a few short-faced bears around, these cats never had a chance to finish off the carcass of their prey. Modern wolf packs and grizzly bears routinely drive mountain lions from their deer and elk kills before the cougar finishes its meal. No sabertoothed cat, which most believe to have been a lonely hunter, could stand up to a short-faced bear or a pack of dire wolves. The American lion was a relatively rare visitor to the tar pits and the sex-structure (2/3 male) might indicate these were wandering, solitary animals. Increased carcass utilization might have been impossible due to competition with those scavengers. La Brea is a spectacular and perhaps unique death trap.

～

In conclusion, the question of the Late Pleistocene extinction of the megafauna begs the entire quandary of the peopling of the Americas.

Climate certainly paves the way for the two-legged creatures (we assume Asian origins). The hardest evidence points to virtually no humans getting south into the lower-48 until the global warming of about 15,000 years ago. Northwest glaciers begin to melt and a coastal route using boats opens along the Pacific. At about the same time, the earliest credible American sites show up. Those few humans, the pre-Clovis people, made little discernible impact on the land or its wildlife: These dozen or so scattered sites indicate a transient people using a non-descript tool kit of small artifacts. There is no evidence they thrived as a culture or population. The climate continues to warm between 14,700 and 12,900 years ago. The impact on the flora and fauna is unknown but shifts in vegetation distribution and habitat fragmentation are expected. Likely the megafauna, those herds of herbivores, migrated widely throughout the changing landscape, their predators close behind. Food stress might be a factor.

The continental ice sheets continue to recede opening a corridor through the ice from eastern Beringia southward to the Rocky Mountain Front of Alberta and northern Montana. By 13,300 years ago, the ice-free corridor is sufficiently revegetated to allow elk to come down from Alaska. Humans, with their dogs, pemmican and waterfowl hunting skills, could have used the same passage earlier than the elk. But there is no sign of such a migration until the time of Clovis, about 13,100 years ago. Why the delay in using the corridor and what kept the earlier pre-Clovis people from flourishing? The presence and abundance of Late Pleistocene predators and carnivores, especially the short-faced bear, are worthy suspects.

Suddenly, Clovis people pop up. Probably out of the corridor but perhaps from elsewhere else. Someone invents or imports the massive, deadly Clovis projectile point and all hunting hell breaks loose. Mammoth show up on the menu. If Clovis people came down through the ice-free corridor, they first encountered mammoth in southern Alberta and probably invented the Clovis point at one of the lithic quarries south of the Missouri River in Montana. A few authorities believe that a separate non-Clovis big-game hunting kit was developed at the same time in the far West. The killing of megafauna begins (some think merely continues) and within two-hundred years the feared Clovis points are found south of the ice at the far corners of North America, some of these fluted blades scattered among or imbedded in the bones of extinct mammals.

About thirty-five genera of mammals disappear from America, about half of them in a brief window of 500 years, 13,200 to 12,700 years ago, with Clovis hunters occupying the core of that time period. A sudden cooling, the Younger Dryas, descends on the Earth by 12,880 years ago, marking the terminal appearance of many of these animals. Suspected causes of the YD are still contentious. But it signals the end of Clovis and much of the megafauna. Thirteen hundred years later, it warms, the Holocene begins and the world is changed. Agriculture sprouts and for the next 12,000 years we coast along the stable climatic highway that brings us all the way to the towering fiery crisis of today.

Though the jury is still out, the rap doesn't look good for *Homo sapiens*. Certainly, a chain of ecological events could have been set off by climate change and by early Americans killing off a few mammoth or sloth, which in turn started a cascading series of destructive habitat changes, even ecosystem collapse, that led to the decline of the megafauna. If hunters remove just 4 or 5% of a population of slow-reproducing wildlife, those animals are on a road headed toward extinction. But, if the causative conditions of gradual decline existed, they're hard to document. It doesn't seem to matter if the extinct animals were browsers, grazers or carnivores (the extinction of predators dependent on declining prey is a simple relationship); they all went down. Some evidence indicates that mammoth herds were expanding their range during this time. This time frame suggests that Clovis hunters had the most blood on their hands. There is little evidence that the small population of pre-Clovis people hunted big animals.

I wish I could find a more satisfying explanation. When I began this book, I thought the evidence leaned towards climate change as the primary cause of the extinction of these magnificent animals, with humans—still a necessary component—merely delivering the coup-de-grace. Hunters certainly didn't knock off all the megafauna animal by animal. And the Younger Dryas is troublesome and suspect: At sites in Arizona, sloth, tapirs, mammoth, horses, mastodons, camels, dire wolves, lions and short-faced bears disappear at the black-mat boundary. Elsewhere in North America, the moment of extinction is less precise. But 17 genera of wildlife get rubbed out in the last days of the Pleistocene, most all in that lethal moment 13,000 to 12,700 years ago. Extinction dates that fit this cluster are still coming in.

The one unmistakable lesson of Late Pleistocene extinction is that human activity combined with global warming is a potential, ageless, deadly blueprint for ecological disaster.

These two phenomena are again at our doorstep today. What will we do? Will we recognize the ancient shadow of that sabertoothed cat in our modern shrubbery and forge a brand new line of defense?

A curious and perhaps unproductive observation is that the amount of time the Clovis people impacted the Earth spanned about 250 years. European colonialists achieved a firm toehold across North America somewhere between two and three centuries ago.

That iconic Clovis projectile point—simple, luminous, so skillfully sculpted that some think it was intended for totemic exchanges in religious ceremonies—was the first in a long line of cold, deadly American weapons, as sleek and beautiful as a Patriot missile.

~

Today, we face change in a magnitude of global warming that will make the extinctions of the Late Pleistocene seem like small potatoes. Not even the grizzly bear will survive the worst-case scenario for today's human-induced climate changes. The fate of the two-legged omnivore in the 22nd century is itself up for grabs. Our perception of what urgently lies in our own vital interests for human survival is barely coming into focus. And it appears to be very late in the climate-change game: Scientists believe we are reaching the "tipping" point in global warming (the polar ice caps melt, then the Amazonian rainforest collapses, which leads to a Siberian permafrost thaw, releasing methane that gives us an increased 6 degree Celsius, etc.), where natural systems experience sudden, rapid and irreversible change.

Scientists are constantly playing catch-up when it comes to underestimating the magnitude and consequences of a warming climate; it's not surprising the public doesn't seem to get it. Governments have no such excuse: They mandate use of Styrofoam cups not reduction of greenhouse gas emissions. And even if they did act, it may be too late for governmental bailouts and technological fixes. The problem is not simply greenhouse gas concentrations; it's a way of life whose time is over. Endless growth is impossible in a world of finite resources.

The hardest truth from the hardest thinkers is that we must walk away from 12,000 years of stable climate and 10,000 years of civilization to find a new way of living. Our version of human life on earth is ending and it's useless to pretend we can hang on to our current lifestyle. We'll have to give up much of what we consider central to our usual daily lives; I'm not looking forward to it and mostly see severe, global cutting-back as a painful and sometimes ugly process. There are no benign economic solutions and corporations, driven by the myth of continual growth, are unlikely to arrive on the doorstep of social responsibility.

Technology is what has empowered us to bring on devastating climatic changes within a mere 250 years, which under natural cycles would have taken hundreds of thousands of years. (Might we make a similar but much smaller claim for the iconic Clovis projectile point contributing to the demise of the megafauna on an identical timetable?) Despite misplaced hope for global sunshades and other temporary fixes to buy some time, high technology will not bail us out of this one.

∿

Will we be left with James Lovelock's predictions? What would be your emotional capacity to endure if you were the only survivor of a family of seven? Six-billion people, Lovelock says, may die off by the end of this century (it could be earlier), the agricultural lands of Africa, Asia and Europe dry up, urban survivors flee the starving, flooded cities and head northward into the new population centers of the Arctic ruled by brutal warlords. The thugs deny them entry and they head out again, leading their bleating camels through the unbearable heat toward the next hot oasis.

∿

We may have heard this speech before: "Growth for the sake of growth is the ideology of a cancer cell."

I thought I buried that kind of talk when three friends and myself tossed Edward Abbey into a beautiful desert grave some 23 years ago, in accordance with his last wishes—illegally of course. Abbey, one of the closest and most influential friends of my life, started telling me this story back in 1968, the year we met, the same year he published his masterpiece, *Desert Solitaire.*

The beasts stalking the edges of our human world today—nuclear warfare and global warming—were already taking shape in the landscape of *Desert Solitaire.* Abbey tells a story wrapped around human rapaciousness for uranium mines. At the end of the tale, a child experiences a beautiful hallucination of the living earth, then dies of massive overexposure to radiation—from atomic rays or rays from the sun, a parable of apocalyptic war or approaching climate change? Abbey provides no easy answers, but this much is evident: Just as we grasp our absolute need for the wild beauty of the world, we are losing it. And it is the children who will pay the price.

Ed is clear about what drives this madness: human greed exemplified by too many people living too high on the hog. "The ever-growing economy," he wrote, "is based on the superstition that we can steal from our children. And without getting caught." Abbey foretells the collapse of industrial civilization, warning us that we must reduce our industrial footprint before catastrophe does it for us. Time and wind will bury the polluted cities of the Southwest, he warns, "growth for the sake of growth is a cancerous madness." Out of this wasteland, the boldest of survivors will wander a new wilderness and perhaps get it right the second time.

That was 1968.

Abbey pops up at the end of this book because I made him a promise fifteen years ago (the fact that he was dead does not relieve me of that obligation). Ed's statement that "wilderness, in America or anywhere else, is the only thing left that is worth saving" parallels this book's contention that the preservation of wild habitats will be central to our own struggle to survive the climatic upheavals of the 21st century. Abbey's predictions are now upon us. Ed might suggest we gather up a like-minded tribe of friends and figure out how to live locally and sustainably (a word Abbey would never use): Stash some beans, hide some peanut butter, and bury lots of ammo and a good deer rifle. Maybe several. Above all, we would have to re-learn the skills of our hunter-gatherer ancestors who managed to survive for a couple of hundred thousand years without agricultural and industrial technology.

Most importantly, Ed would tell us to fight like hell, here, right now: It is a patriot's duty to defend his planet against the corporations and the governments that work for them, who are poisoning the health of this precious, fragile blue orb we live on. For now, get active, eat low on the food chain, walk don't drive, stay home, get green, unplug—do the small things. And always, fight to protect the wild; wilderness will buy us the kind of time geo-engineering never could or will. But it might be very late in the climate game to plan for transition. The physics of climate change will shrink our options and close many doors to us. And then it will be time for the survivors to re-enter the remnants of wild landscapes from whence we came, that original homeland that carved the human mind at the beginning of our kind's time, when living was beautiful, vivid and wondrous, dangerous animals riveted our attention.

It's time to confront our own sabertoothed cat. What will we do? Will we be bold, adaptive, sufficiently opportunistic and adventurous to successfully face the great climatic shifts of the 21st century? The question may appear less argument than allegory but it is far less esoteric than it sounds. Those ancient mammoth hunters found a way to carry on. If only for a while. May we be half as lucky?

Acknowledgements

Guggenheim and Lannan foundations for their financial support.

Dennis Sizemore, Rick Bass, Chris Filardi, Peter Gerity, Doug Milek, Michael Soulé, Pic Walker, John Ward, and Terry Tempest Williams from Round River Conservation Studies.

To the elders of the tribe, Peter Matthiessen and Jim Harrison.

Archeologist Larry Lahren, anthropologist Bernard "Bunny" Fontana and biologist Lloyd Findley for lending their time and expertise.

Archeologist Mark Papworth, who started me on this path, and his widow, the lovely Linda, for putting me up (and putting up with me) all these years. To Jeff and Susan Bridges for the same reason.

Outside magazine for printing the original Anzick site story.

Matt Smith for letting me fondle his short-faced bear skeleton.

Ben Hargis, Calvin Sarver and Helen Anzick families for their generosity in sharing their stories.

Friends Pat Armstrong, Marc Beaudin, Joe Camp, Dan Drislane, Donna Greenberg, Loren Haarr, Fenia and Carl Hiaasen, Margot Kidder, Bud Lucas, Craig Mielke, David and Betsy Quammen, Dan and Carole Sullivan.

Professors Mark Aronson, David Cremean, John Doris, Greg Keeler, and David Price for their tireless enthusiasm for teaching.

Chuck Irestone, who still won't take a penny for his technical support.

All of AK Press and CounterPunch: Tiffany Wardle, Joshua Frank, Christa B. Daring, Zach Blue, Jeffrey St. Clair, Becky Grant, Lorna Vetters and especially the late, great Alex Cockburn.

Phyllis, Laurel and Colin Peacock.

Bibliography

Chapter 2   The Lair of the Short-faced Bear

Dunham, Will. "Arctic ice sheet collapse may swamp U.S. coasts." *Reuters,* 5 February 2009.

Figueirido, Borja, Juan A. Pérez-Claros, Vanessa Torregrosa, Alberto Martín-Serra and Paul Palmqvist. "Demythologizing Arctodus simus, the 'Short-Faced' long-legged and predaceous bear that never was." *Journal of Vertebrate Paleontology* Vol. 30, No. 1, 2010: 262–275.

Geist, V. "Large predators: them and us!" *Fair Chase* Vol. 23, No. 3, 2008: 14–19.

Goodell, Jeff. "The Prophet of Climate Change: James Lovelock." *Rolling Stone,* 17 October 2007.

Hotz, Robert Lee. "Greenland's ice sheet is slip-sliding away." *Los Angeles Times,* 25 June 2006.

Matheus, Paul E., "Diet and co-ecology of Pleistocene short-faced bears and brown bears in Eastern Beringia." *Quaternary Research*, Vol. 44, 1995: 447–453.

—. "Locomotor adaptations and Ecomorphology of short-faced bears (*Arctodus simus*) in eastern Beringia." Yukon Paleontology Program, *Occasional papers in earth sciences,* No. 7, 2003.

—. "Paleoecology and ecomorphology of the giant short-faced bear in eastern Beringia." Unpublished Ph.D. Dissertation, University of Alaska, Fairbanks, 1997.

Revkin, Andrew C. "Updating Prescriptions for Avoiding World Wide Catastrophe: A Conversation with James Lovelock." *The New York Times,* http://www.nytimes.com/2006/09/12/science/earth/12conv.html?_r=0 accessed, 23 January 2013.

Schubert, B. W. "Late Quaternary chronology and extinction of North American giant short-faced bears (*Arctodus simus*)." *Quaternary International* 217.1-2, 15 April 2010: 188–194.

Sorkin, B. "Ecomorphology of the giant short-faced bears *Agriotherium* and *Arctodus*." *Historical Biology* Vol. 18, No. 1, March 2006: 1–20.

Vaillant, John. *The Tiger: A True Story of Vengeance and Survival.* New York: Random House, 2010.

Chapter 3 Archaeology and the Shape of the Journey

Goebel, T., Waters, M.R. and O'Rourke, D.H. "The Late Pleistocene dispersal of modern humans in the Americas." *Science* Vol. 319, No. 5869, 2008: 1497–1502.

Haynes, C. Vance Jr. "Clovis Progenitors: From Swan Point, Alaska to Anzick Site, Montana in Less than a Decade." University of Arizona, 2004.

Pitulko, V. V. , P. A. Nikolsky, E. Yu. Girya, A. E. Basilyan, V. E. Tumskoy, S. A. Koulakov, S. N., Astakhov, E. Yu. Pavlova and M. A. Anisimov. "The Yana RHS Site: Humans in the Arctic Before the Last Glacial Maximum." *Science*, 2 January 2004: Vol. 303, No. 5654, 52–56.

Chapter 4 Invisible People

Baharak Hooshiar Kashani, Ugo A. Perego, Anna Olivieri, Norman Angerhofer, Francesca Gandini, Valeria Carossa, Hovirag Lancioni, Ornella Semino, Scott R. Woodward, Alessandro Achilli, Antonio Torroni. "Mitochondrial haplogroup C4c: A rare lineage entering America through the ice-free corridor?" *American Journal of Physical Anthropology,* Vol. 147, Issue 1, 35–39, January 2012.

Cinq-Mars, J., and R. E. Morlan. "Bluefish Caves and Old Crow Basin: A New Rapport." In *Ice Age People of North America: Environments, Origins, and Adaptations*, ed. R. Bonnichsen and K. L. Turnmire, 200. Corvallis: Oregon State University Press, 1999.

Hoffecker, John F. and Scott A. Elias. *Human Ecology of Beringia*. New York: Columbia University Press, 2007.

Matheus, Paul, James Burns, Jaco Weinstock and Michael Hofreiter. "Pleistocene brown bears in the mid-continent of North America." *Science*, Vol. 306, No. 5699, 12 November 2004: 1150.

Meltzer, David J. *First Peoples in a New World: Colonizing Ice Age America*. Berkeley: University of California Press, 2009, 155–170.

Nelson, Richard K. *Hunters of the Northern Forest: Designs for Survival among the Alaskan Kutchin*. University of Chicago Press, 1986.

Pavlov, Pavel, John Inge Svendsen and Svein Indrelid. "Human Presence in the European Arctic nearly 40,000 years ago." *Nature,* No. 413, 6 September 2001: 64–67.

Pitulko, V. V. , P. A. Nikolsky, E. Yu. Girya, A. E. Basilyan, V. E. Tumskoy, S. A. Koulakov, S. N., Astakhov, E. Yu. Pavlova and M. A. Anisimov. "The Yana RHS Site: Humans in the Arctic Before the Last Glacial Maximum." *Science*, 2 January 2004: Vol. 303, No. 5654, 52–56.

Straus, Lawrence Guy. "Solutrean settlement of North America? A review of reality." *American Antiquity,* Vol. 65, No. 2, April 2000: 219–226.

Tamm, Erika, Kivisild, Toomas; Reidla, Maere; Metspalu, Mait; Smith, David Glenn; Mulligan, Connie J.; Bravi, Claudio M.; Rickards, Olga; Martinez-Labarga, Cristina; Khusnutdinova, Elsa K.; Fedorova, Sardana A.; Golubenko, Maria V.; Stepanov, Vadim A.; Gubina, Marina A.; Zhadanov, Sergey I.; Ossipova, Ludmila P.; Damba, Larisa; Voevoda, Mikhail I.; Dipierri, Jose E.; Villems, Richard; Malhi, Ripan S., 2007. Carter, Dee. ed. "Beringian Standstill and Spread of Native American Founders". *PLoS ONE,* Vol. 2, Issue 9, 829.

Chapter 5   Mingled Fates of Homo sapiens and Ursus arctos horribilis

Barnes, I., P. Matheus, B. Shapiro, D. Lowell Jenson and A. Cooper. "Dynamics of Pleistocene population extinctions in Beringian brown bears." Science Vol. 295, No. 5563, 22 March 2002: 2267–2270.

Leonard J. A., Wayne, R. K. and Cooper, A. "Population genetics of ice age brown bears." Department of Organismic Biology, University of California, Los Angeles, CA 90095-1606, USA. *Proceedings of the National Academy of Sciences of the United States of America*, 15 February 2000; Vol. 97, No. 4: 1651–1654.

Morrow, J. and S. J. Fiedel. "New radiocarbon dates for the Clovis component of the Anzick site, Park County Montana." *In Paleoindian Archaeology: A Hemispheric Perspective*, ed. J. Morrow and C. Gnecco. Gainesville: University Press of Florida, 2007.

Chapter 6   Braving the Northwest Coast During the Time of Icebergs

Arnold, Thomas G. "Radiocarbon dates from the ice-free corridor." *Radiocarbon* Vol. 44, No. 2, 2002: 437–454.

"Broken Mammoth Archaeological Project, Introduction." Alaska Department of Natural Resources, Internet. 01 April 2011.

Dixon, E. J. *Bones, Boats, and Bison: Archeology and the First Colonization of Western North America*. Albuquerque: University of New Mexico Press, 1999.

Gualtieri, Lyn. "Testing the sensitivity of two 36Cl age calculation programs" Ph.D. Thesis, University of Massachusetts. Reported 2000.

Jackson, Lionel E. Jr., Phillips Fred M., Shimamura, Kazuharu and Little, Edward C. "Cosmogenic 36Cl dating of the Foothill Erratic Trains, Alberta, Canada." *Geology,* Vol. 25, No. 3: 195–198.

Jackson, Lionel E. Jr. and Wilson Michael C. "The Ice-Free Corridor Revisited." *Geotimes,* February 2004.

Marean, Curtis W. "When the Sea Saved Humanity." *Scientific American,* Vol. 303, No. 2, August 2010.

Meltzer, David J. *First Peoples in a New World: Colonizing Ice Age America.* Berkeley: University of California Press, 2009.

Wong, Kate "The Morning of the Modern Mind." *Scientific American*, June 2005.

Chapter 7 Pre-Clovis People

Adovasio, J. M. and Page, J. *The First Americans: In Pursuit of Archaeology's Greatest Mystery.* Modern Library, 2002.

Adovasio, J. M., D. Pedler, J. Donahue and R. Stuckenrath. "No vestige of a beginning nor prospect for an end: two decades of debate on Meadowcroft rockshelter." In *Ice Age People of North America: Environments, Origins, and Adaptations,* ed. R. Bonnichsen and K. L. Turnmire, 416–431. Corvallis: Oregon State University Press, 1999.

Collins, M. B. "The Gault Site, Texas, and Clovis Research." *Athena Review* Vol. 3, No. 2, 2002: 24–36.

Dillehay, T. D. *Monte Verde: A Late Pleistocene Settlement in Chile.* Vol. 2, *The Archaeological Context and Interpretation.* Washington, D.C. Smithsonian Institution Press. Washington D.C., 1997.

Dillehay, T.D., C. Ramirez, M. Pino, M.B. Collins, J. Rossen, and J.D. Pino-Navarro. "Monte Verde: seaweed, food, medicine, and the peopling of South America." *Science,* Vol. 320, No. 5877, 2008: 784–786.

Gilbert, M.T.P., D.L. Jenkins, D.L., A. Gotherstrom, N. Naveran, J.J. Sanchez, M. Hofreiter, P.F. Thomsen, J. Binladen, T.F.G. Higham, R.M. Yohe, II, R. Parr, L.S. Cummings and E. Willerslev. "DNA from pre-Clovis human coprolites in Oregon, North America." *Science,* Vol. 320, No. 5877, 2008: 786–789.

Harmon, Katherine. "People Were Chipping Stone Tools in Texas More Than 15,000 Years Ago." *Scientific American*, 24 March 2011.

Jenkins, D.L., L.G. Davis, T.W. Stafford Jr., P.F. Campos, B. Hockett, G.T. Jones, L.S. Cummings, C. Yost, T.J. Connolly, R.M. Yohe II, S.C. Gibbons, M. Raghavan, M. Rasmussen, J.L.A. Paijmans, M. Hofreiter, B.M. Kemp, J.L. Barta, C. Monroe, M.T.P. Gilbert, E. Willerslev. "Clovis age Western stemmed projectile points and human coprolites at the Paisley Caves." *Science,* Vol. 337, No. 6091, 2012: 223–228.

Lawler, A. "Pre-Clovis mastodon hunters make a point." *Science* Vol. 334, No. 6054, 2011: 302.

McAvoy, J. M. and L. D. McAvoy. "Archaeological Investigations of Site 44sx202, Cactus Hill, Sussex County, Virginia." *Nottoway River Survey Archaeological*

Research Report No. 2 and Research Report Series No. 8. Richmond, VA: Department of Historic Resources, 1997.

Morgan, Patrick. "New archeological find buries theory on First Americans, re-opening a gaping mystery." *Discover Magazine*. 3 March 2011.

Redmond, Brian, McDonald, Greenfield, H. J. and Burr, M. L.,"New evidence for Late Pleistocene human exploitation of Jefferson's Ground Sloth (Megalonyx jeffersonii) from northern Ohio, USA." *Journal: World Archaeology*, Vol. 44, No. 1, 75–101, 2012.

"Schafer and Hebior Mammoth Sites." Friends of the Ice Age. Kenosha Public Museum, Kenosha County, Wisconsin.

Topper. "New Evidence Puts Man In North America 50,000 Years Ago". *Science Daily*. 18 November 2004.

Waters, Michael R., Steven L. Forman, Thomas A. Jennings, Lee C. Nordt, Steven G. Driese, Joshua M. Feinberg, Joshua L. Keene, Jessi Halligan, Anna Linquist, James Pierson, Charles T. Hallmark, Michael B. Collins and James E. Wiederhold. "The Buttermilk Creek Complex and the Origins of Clovis at the Debra L. Friedkin Site, Texas." *Science,* Vol. 331, No. 6024, 25 March 2011: 1599–1603.

Chapter 8 Clovis

Collins, M. B. "The Gault Site, Texas, and Clovis Research." *Athena Review,* Vol. 3, No. 2, 2002: 24–36.

Haynes, C. Vance Jr. and Bruce B. Huckell. *Murray Springs: A Clovis Site with Multiple Activity Areas in the San Pedro Valley, Arizona.* Tucson: University of Arizona Press, 2007.

Lahren, L. *Homeland: An archeologist's view of Yellowstone Country's past.* Cayuse Press, 2006.

Lahren, L. and R. Bonnichsen. "Bone Foreshafts from a Clovis Burial in Southwestern Montana." *Science,* Vol. 186, No. 4159, 1974: 147–150.

Morrow, J. and S. J. Fiedel. "New radiocarbon dates for the Clovis component of the Anzick site, Park County Montana." In *Paleoindian Archaeology: A Hemispheric Perspective,* ed. J. Morrow and C. Gnecco. Gainesville: University Press of Florida, 2007.

O'Brien, Michael and Wood ,W. Raymond. *The Prehistory of Missouri.* Columbia. University of Missouri Press, 1998.

Owsley, Douglas W. and David R. Hunt. "Clovis and Early Archaic Crania from the Anzick Site (24PA506), Park County, Montana." *Plains Anthropologist,* Vol. 46, No. 176, May 2001: 115–124.

Stanford, Dennis J. and Bruce A. Bradley. *Across Atlantic Ice: The Origins of America's Clovis Culture.* Berkeley: University of California Press, 2012.

Waters, Michael L. and Thomas W. Stafford Jr. "Redefining the Age of Clovis: Implications for the Peopling of the Americas." *Science,* Vol. 315, No. 5815, 23 February 2007: 1122–1126.

Chapter 9 Endgame

"13,000-Year-Old Stone Tool Cache in Colorado Shows Evidence of Camel, Horse Butchering." University of Colorado. 25 February 2009.

DeSantis, LRG, Schubert, BW, Scott, JR, Ungar PS, 2012. "Implications of Diet for the Extinction of Sabertoothed Cats and American Lions." *PLoS ONE,* Vol. 7, No. 12: 52453.

Fiedel, Stuart and Gary Haynes. "A premature burial: comments on Grayson and Meltzer's 'Requiem for overkill.'" *Journal of Archaeological Science,* Vol. 31, 2004: 121–131.

Firestone, R.B., A. West, J.P. Kennett, L. Becker, T.E. Bunch, Z.S. Revay, P.H. Schultz, T. Belgya, D.J. Kennett, J.M. Erlandson, O.J. Dickinson, A.C. Goodyear, R.S. Harris, G.A. Howard, J.B. Kloosterman, P. Lechler, P.A. Mayewski, J. Montgomery, R. Oreda, T. Darrah, S.S. Que Hee, A.R. Smith, A. Stich, W. Topping, J.H. Wittke and W.S. Wolbach. "Evidence for an extraterrestrial impact 12,900 years ago that contributed to the megafaunal extinctions and the Younger Dryas cooling." *Proceedings of the National Academy of Sciences,* Vol. 104, No. 41, 9 October 2007: 16016–16021.

Gill, Jacquelyn L., Williams, John W., Jackson, Stephen T., Lininger, Katherine B. and Robinson, Guy S. "Pleistocene Megafauna Collapse, Novel Plant Communities, and Enhanced Fire Regimes in North America." *Science,* 20 November 2009.

Graysona, Donald K. and David J. Meltzer. "A requiem for North American overkill" *Journal of Archaeological Science* 30, 2003, 585–593.

Haynes, Vance Jr. "The Rancholabrean Termination." In *Paleoindian Archaeology: a Hemispheric Perspective,* ed. J. Morrow and C. Gnecco. Gainesville: University Press of Florida, 2007.

Kelly, R.L. and L. C. Todd. "Coming into the country: Early Paleoindian hunting and mobility." *American Antiquity,* Vol. 53, No. 2, 1988: 231–244.

Martin, P. S. and R. G. Klein, eds. *Quaternary Extinctions, a Prehistoric Revolution.* Tucson: University of Arizona Press, 1984.

Raper, Diana and Mark Bush. "A test of Sporormiella representation as a predictor of megaherbivore presence and abundance." *Quaternary Research,* Vol. 71, No. 3, May 2009: 490–496.

Robinson, Guy S., Burney, Lida Pigott and Burney, David A. "Landscape Paleoecology and Megafaunal Extinction in Southeastern New York State" *Ecological Monographs,* "Evidence for an extraterrestrial impact 12,900 years ago that contributed to the megafaunal extinctions and the Younger Dryas cooling." "Evidence for an extraterrestrial impact 12,900 years ago that contributed to the megafaunal extinctions and the Younger Dryas cooling." Vol. 75, No. 3, August 2005, 295–315. Published by Ecological Society of America.

Index

Abalone, 125, 130, 139
Abbey, Edward, 159, 202
Ablation, glacial, 27
Absaroka Range, 14–5
Accelerator Mass
  Spectrometry, 151
Accumulation, glacial, 149
Across Atlantic Ice (Stanford)
  175, 177
Admiralty Island, 113
Africa, 19, 38, 42, 52, 81–2,
  124–26, 168–69, 185, 201
Agave, 108–9
Agriculture, 34, 38, 186, 191,
  199
AIDS, 191
Alaska, 1–2, 6–7, 14, 24, 38–9,
  66–7, 72, 78, 93, 97, 112–13,
  116–17, 119–20, 174, 184, 190
Alberta, Canada, 72, 98, 114,
  174, 195, 198
Aleut people, 18, 92
Aleutian Islands, 83
Alexander Archipelago, 133
Altai Mountains (Mongolia),
  56
Amanita mushrooms, 59–60
Amazon, 200
Amba River (Siberia), 44,
  46–7
Andes Mountains, 84
Arctica, 36, 81
Anthropology, 124–25, 148
Anzick site (Montana), 9–18,
  20, 75, 87, 117, 147, 166–67,
  170–82, 187
Anzick, Helen, 12, 172
Anzick, Mel, 9–10, 12–3
Anzick, Sarah, 13, 182
Appleman Lake (Indiana), 194
Archaeology, 3–4, 6–7, 9–18,
  20–1, 33–5, 42, 63–5, 110–11,
  147–48, 165–66, 180–81
Archaic culture, 6, 123

Arctic Ocean, 6, 85, 115,
  119–20
Arctic, 17, 33, 43, 65, 81–2,
  85–8, 128, 132, 153–55, 162,
  201
Arizona, 6, 120, 148, 187
Arkansas State University,
  182–83
Armagosa points, 123
Aronson, Mark, 171
Arrowheads, 4–5, 7, 21, 73, 165,
  172, 175
Artifacts, 4–5, 7, 9–10, 12–14,
  24, 33, 51, 56–7, 68–9, 71–3,
  75, 89, 137, 145, 149–51, 157,
  165–68, 170–72, 175, 180–81,
  198–200
Asteroid impact, 38, 186,
  192–93
Astronomy, 123–24
Athapaska people, 101
Atlantic Ocean, 37–8, 85
Australia, 2, 41, 82–3, 104,
  127, 137
Aztec culture, 170
Baja Peninsula, 120
Baranof Island, 113
Beluga whale, 153–54
Bering Strait, 2, 28–9, 43, 58,
  64, 66, 81–3, 85–6, 90–1, 150,
  187–88
Beringia, 8–11, 28, 39, 43, 68,
  72, 81, 90, 93, 102, 105, 107,
  112–3, 127, 145, 161, 174, 193
Beringian Standstill theory,
  91, 97–8
Bifaces, 10, 56, 71, 89, 151,
  170–72, 180–81
Bighorn sheep, 109
Bikin River (Siberia), 44–8
Binford, Louis, 6
Biophilia, 19
Bison, American, 30, 64–5, 78,
  94, 109, 111, 167, 169, 193
Black bear, 47, 111, 129, 167
Black Mat formation, 187–89,
  191–93, 199
Blackfeet Nation, 144–45
Blue Fish Caves site (Yukon),
  100–2
Blue Fish River (Yukon), 100
Blue whale, 141
Boats, 134–38, 139–42

Bølling-Allerød interstadial
  period, 36–7, 93, 189–90,
  196–96
Bonnichsen, Rob, 173–74,
  181–83
Boulder, Colorado, 190
Bradley, Bruce, 175
Bray, Bill 172
Brazil, 102
British Columbia, 7–8, 25–6,
  36–7, 79, 107
Brokaw, Tom, 45–8
Brooks Range (Alaska), 6
Bullboats, 139–42
Burials (Paleoindian), 9–18,
  20, 75, 87, 117, 147, 166–67,
  171–72, 174–80
Burke Channel, 130–31
Butchering sites, 31, 55–6, 85,
  87, 89, 101–3, 141, 148, 151–52,
  162, 172
Buttermilk Creek complex
  (Texas), 155–57, 183
Cactus Hill site (Virginia),
  150–51
California, 32, 109, 137, 142–43,
  194
Camel (Pleistocene), 101, 167,
  187, 190, 201
Canada goose, 4
Canada, 7, 67–8, 78, 114,
  127–28, 192, 195, 198
Candlefish, 129
Canoes, 142
Cape Caution, 128
Carbon-14 dating, 23–4, 73,
  102–3, 106, 113, 133, 143–44,
  149–50, 173–74, 180–81, 183
Caribou, 78, 109, 141, 154, 193
Cascade Mountains, 158
Catfish, 125
Cave bears, 105
Center for the Study of First
  Americans, 182–83
Channel Islands, 137
Chapala Basin site (Mexico),
  84
Charcoal remains, 84, 133, 143,
  148–50
Chert, 4–5, 10, 86, 156–57,
  172, 190
Chesapeake Bay, 176
Cheyenne Nation, 182

Chichagof Island, 113
Chihuahua, Mexico, 48, 79
Chile, 66, 70, 84, 104, 136, 143, 148–50, 156, 161
China, 93–4, 103
Chippewa Nation, 5
Chouinard, Yvon, 45–8
Clams, 131–32, 139
Climate change, 1–3, 15–6, 22, 33–4, 37–8, 64–5, 77–8, 81, 117, 124–28, 148, 165–67, 184–203
Clovis culture, 2–3, 6–7, 9–18, 20–1, 30–1, 38, 40–1, 50–1, 53, 55–60, 63–4, 87, 91, 99–100, 117, 133, 143, 147–48, 149–50, 165–82, 185–86, 198–200
Clovis First theory, 50, 53, 69–71, 148, 157, 160, 174
Clovis point, 9, 12–14, 24, 33, 51, 56–7, 68–9, 72, 75, 137, 145, 149–50, 157, 165–68, 198–200
Cognitive (also Behavioral) modernity theory, 124–126
Colorado River, 18
Colorado, 115–16, 190, 194
Columbia River, 75–6, 128, 142, 152, 161
Comet impact theory, 192–93
Continental Divide (N. America), 26
Coprolites, 148, 158–63
Cordilleran ice sheet, 17–8, 30, 67, 99–100, 129
Coyote, 159–60
Crabs, 122, 130–32, 136
Crater Lake, 158
Crow Nation, 182
CT scanning, 149
Deglaciation, 1–2, 16, 25, 27, 33, 36, 38–9, 65, 68–9, 100, 117, 131–32, 136, 143, 193, 197, 200
Denmark, 177
Dental evidence, 92
Deschutes River, 161
Desert Solitaire (Abbey), 202
Dhole dog, 32
Dire wolf, 1, 30, 60, 84, 97, 141, 187, 197, 199
Discover magazine, 156–57

Discovery channel, 74
DNA testing (ancient), 13, 91, 98, 106, 113, 149, 151, 158–63, 177, 180–82, 192
Dodo, 86
Dogs, 32, 93–4, 144, 152, 155, 186, 189, 191, 198
Drought, 126–27
Dung fungus, 194–95
Dust, study of, 63
East China Sea, 83
Economics, 22
Edmonton, Alberta, 114
Edwards Plateau (Texas), 157
Elephants, 168–69
Elk, 109, 111, 117, 172–74, 187, 193, 198
Ellison, Jib, 45
Erratic rocks (glacial), 28, 144–45, 174
Evolution (human), 77, 124–25
Extinction, 3, 16–7, 40, 63, 74, 78, 115, 165, 168–69, 185–86, 188–89
Famine, 126–27
Fin whale, 121, 141
Fingerprints, of paleoindians, 86
Fire, 84, 93–6, 105, 133, 150–51
Fitzhugh Sound, 131
Flathead Creek (Montana), 172
Florida, 55, 70
Folsom culture, 188, 191
Fontana, Bernard, 8
Food stress theory, 198–99
Footprints, of paleoindians, 86
Forests, 39–40, 103, 140–41
Fort Yukon, 100
Fossil Lake (Oregon), 161
Fossils, 30, 55, 68, 101–2, 112–13, 133, 143
Fox, 159
France, 71, 85, 166
Frasier River, 152
Friedkin site (TX), 155–57
Gaia Hypothesis, 34–5
Gardiner Post Office (Montana), 179
Gathering, 93–5, 108–9
Gault site (Texas), 155–57, 183
Geist, Valerius, 40–1, 55–6

Genetics, 21, 66, 91
Genocide, 22
Geology, 145
Geology, 63, 75, 78, 144
Glacial melt, 1–2, 16, 25, 27, 33, 36, 38–9, 65, 68–9, 100, 117, 131–32, 136, 143, 193, 197, 200
Glacier National Park, 26–7, 48, 113, 136
Glaciers, 16, 24, 26–7, 33–5, 39–40, 65, 67–8, 77, 116–17, 134, 144
Glaciology, 21
Global cooling, 37–8
Global warming, 1–3, 15–6, 22, 33–4, 37–8, 64–5, 77–8, 81, 117, 124–28, 148, 165–67, 184–203
Glyptodont, 31
Gold mining, 72, 97
Gomphothere, 167–68
Grayling, 101
Great auk, 87
Great Falls of the Missouri, 88
Great Lakes, 4, 6, 38, 186, 192
Great Plains, 72, 109, 169, 186
Green Beret (US Army), 60, 120
Greenhouse gasses, 126–27
Greenland, 28, 36–7, 153, 187
Griffin, James B., 6
Grizzly bear, 7–8, 15, 41–3, 48–9, 51–3, 72, 78–9, 84, 88, 94, 98–100, 105–117, 128, 133, 139, 153, 160, 193–94, 197, 200
Grizzly Years (Peacock), 8, 11
Ground sloth, 31, 152, 190, 199
Guggenheim fellowship, 21
Gulf of California, 120–22
Guthrie, A.B., 186
Gyrfalcon Lake (Montana), 27
Haida Nation, 130
Hargis, Ben, 10, 180
Harpoons, 125
Harrison, Jim, 9–10
Haynes, C. Vance, 148
Hearths, 150–51
Hecate Strait, 136
Heilstuk Nation, 127–28, 130
Himalayan Mountains, 25–6
Hinton, Thomas, 8
HIV, 191

Holocene epoch, 37–8, 195, 199
Horse, (Pleistocene), 65, 101, 167, 187, 190, 199
Hunting, 86–9, 108–9, 115–16, 139–42, 144, 166–70, 189–91
Hypothermia, 135
Hyundai, 45–6
Iberian Peninsula, 71, 84–5
Ice Sheets, 27–8, 30, 36–7, 67, 99–100
Ice-Free Corridor (IFC), 9, 28–9, 39, 67–8, 72–3, 92, 99–100, 113–14, 117, 143, 145–46, 158, 160–61, 166, 173–74, 187
Idaho Accelerator Center, 175
Idaho, 175, 178
Indiana, 194
Inuit people (Eskimo), 7, 18, 152–53, 155
Iowa, 171–72
Island biogeographic theory, 188–89
Jaguarundi, 121–22
Japan, 82–3
Javalina, 121–22
Jewelry, 125
Kamchatka Peninsula, 83
Kansas, 57, 103
Katabatic winds, 100
Kennewick Man, 75–6
KGB, 46
Kimmswick site (Missouri), 168–69
Kurten, Bjorn, 50–1
Kutchin people, 100–1
Kwakiutl Nation, 130
La Brea Tar Pits (California), 32, 197
Lahren, Larry, 9–18, 172–74, 181–84
Lake Agassiz, 37–8
Lake Baikal (Siberia), 17–8, 144
Lake Superior, 4
Lannan Foundation, 21
Last Glacial Maximum (LGM), 28–9, 38–9, 55, 67–8, 73, 81, 85, 90, 106, 114, 126, 145, 150
Laurentide ice sheet, 28–9, 30, 67

Leakey, Louis, 86
Lewis and Clark Expedition, 18, 88
Lewis, Bart, 153–55
Libertad, Sonora, 120
Linguistics, 21, 63, 66–7, 125
Lion, American (Pleistocene), 15, 41, 84, 89–90, 94, 97, 152, 163, 187, 196–97, 199
Lithics, 9, 12–14, 24, 33, 51, 56–7, 68–9, 71–2, 75, 85, 87, 123, 131, 145, 149–50, 156–57, 165–68, 175–76, 178–79, 189–90, 198–200
Livingston Natural History Museum (Montana), 9
Lodgepole pine, 15
Logging industry, 45–6
Los Angeles Times, 35–6
Lovelock, James, 34–5, 201
Lower Sonoran Desert, 121–22
Mackenzie River, 6, 8, 100
Mammoth, 30–1, 41, 69–70, 78, 87–8, 94, 148, 165, 167–68, 184, 187–89, 195, 198–99
Manifest Destiny (doctrine), 19
Maritime economy, 130–37, 162
Mastodon, 30–1, 141–42, 151–52, 167–68, 183, 187, 196, 199
Mauritius, 86
Mayflower (ship), 19
Meadowcroft Rockshelter site (Penn.), 148–50, 158, 184
Megafauna (of Pleistocene), 1–2, 17, 24–7, 40–1, 63, 86–7, 107, 148, 165, 168, 185–86, 189, 198–200
Meltwater lakes, 36, 100, 117, 143
Mexico, 7–8, 30–1, 48, 66, 70, 79, 84, 86, 102, 108–9, 120–22, 167–68
Michigan, 4–6
Microdepitage, 156–57
Middens, 123–24
Middle East, 38
Migration routes, of paleo-indians, 9, 28–9, 39, 67–70, 72–3, 82–4, 71–3, 91–92, 99–100, 104, 113–14, 117,

119–46, 149, 158, 160–61, 166, 173–74, 187
Minke whale, 141
Mississippi River, 55, 109, 168
Missouri River, 18, 70, 88, 167–69, 174, 179
Moa, 87
Mongolia, 56
Montana, 9–10, 18, 26, 48, 134, 185–86, 198
Monte Verde site (Chile), 84, 133, 143, 148–50, 157
Moose, 109, 193
Moraines, 39, 117, 133, 144
Moss agate, 175–76
Mount Toba (Sumatra), 82, 126–27
Mountain goat, 109
Mountain lion (cougar), 112, 197
Mountain pine beetle, 15
Muir, John, 19
Murray Hotel (Montana), 9, 11
Mushrooms, 59–60, 193–94
Musk ox, 31, 78, 94, 116, 154, 193
Mussels, 131–32, 139
Na-Dene language group, 18, 92
Namu, BC, 131
Narwhal, 154
National Geographic, 11–2, 63, 73–4
National Human Genome Research Institute, 13
Native Americans, 5, 7, 18, 76, 92, 100, 120–22, 127–30, 144–45, 152–55, 182
Natural History Museum of Los Angeles, 32
Neanderthals, 82, 105, 124–25
Nevada, 151–52
New Mexico, 57, 86
New York, 194–95
New Yorker, 63, 73
New Zealand, 87, 136
Newfoundland, 107
Nile River, 125
Nootka Nation, 130
North American Graves Preservation and Repatriation Act

(NAGPRA) 76–7, 180, 182–83
North Cascades National Park, 134
Northern Arizona University, 181
Norway, 107, 136
Nottoway River, 150
Nova (PBS), 74
Obsidian, 178
Ochre (red), 9, 89, 125, 172–74, 180–1
Ohio, 28
Oil industry, 6
Old Crow (Yukon), 100
Olduvai Gorge (Tanzania), 86
Olympic Peninsula site, 142, 151–52, 183
Optically Stimulated Luminescence (OSL), 156–57
Orca, 141
Oregon State University, 181
Oregon, 66, 70, 104, 108, 117, 142, 148, 157–63
Overkill theory, 168–69
Oysters, 129, 139
Pacific coastal migration route, 68–70, 72–3, 83–4, 91–2, 104, 119–46, 149, 160–61, 166
Pacific Ocean, 68–70, 73, 82–4, 91–92, 119–20, 158, 160–61, 166
Paisley Cave (Ore.), 157–63
Paleontology, 21, 52–3, 63, 72, 106, 110–11
Palynology, 63
Panama, 70, 161
Papworth, Mark, 6, 8, 11–12, 172, 178, 181
Paradise Valley (Montana), 9
Paralytic Shellfish Disease (PSD), 58–9, 131–32
Parry Channel, 132
Pedra Furada site (South America), 84
Pelican Valley (Yellowstone), 49
Pemmican, 117, 129, 170, 198
Pennsylvania, 68, 103, 150, 162, 184, 192
Perception of risk theory, 3. 20, 77

Peregrine falcon, 100
Phosphoria, 175
Pikunov, Dmitri, 44–8
Pine River (Michigan), 5
Plant collecting, 18, 40, 57–59, 65, 75, 78, 93–5, 99, 107–9, 119, 121, 123, 170, 190
Pleistocene epoch, 1–2, 6–7, 13, 16, 25–7, 49, 53–4, 77–8, 105–6, 119–20, 148, 174, 186–87
Point Barrow, Alaska, 119–20
Poison Cove (BC), 131
Poisonous plants, 58–60
Polar bear, 152–55
Polar ice caps, 27
Pollen studies, 143, 194
Porcellite, 175
Porcupine River (Alaska) 8, 100–1
Pottery, 123
Powell, John Wesley, 18
Pre-Clovis cultures, 24, 42–3, 55, 66, 70, 70–1, 74, 84–6, 133, 147–63, 166–67
Pre-Max glacial period, 91–2, 95, 98–100, 102, 116, 127
Prince of Wales Island (Alaska), 116–17
Psychedelic plants, 59–60
Puffins, 139
Quarries, 70, 167, 175–76, 178–79
Queen Charlotte Channel, 136
Queen Charlotte Islands, 8, 131
Queen Charlotte Strait, 128
Racism, 76
Radiocarbon dating, 23–4, 73, 102–3, 106, 113, 133, 143–44, 149–50, 173–74, 180–81, 183
Red Sea, 127, 136
Red tides, 58
Reindeer, 65
Repatriation, 1, 8, 76, 180, 183
Rewilding movement, 186
Ridgeway, Rick, 45–8, 153–55
RNA testing (ancient), 192
Rocky Mountain Trench, 144
Rocky Mountains, 6–9, 26, 28, 33, 70, 72, 79, 120, 144–45, 178, 198
Ross Ice Sheet, 36–7

Round River Conservation Studies, 127–28
Rummels-Maske site (Iowa), 171–72
Russia, 8, 14, 17–8, 24, 38–9, 64–6, 72, 81–2
Ryukyu Islands (Japan), 82–3
Sabretooth cat, 1, 7, 32–3, 60, 78, 84, 143, 196–97, 203
Saginaw Basin (Michigan), 6
Saiga antelope, 94
Salmon, 101, 109–10, 129–30, 132, 141–42, 162
Saltville site (Virginia), 103, 151
San Carlos, Sonora, 120
San Juan Islands, 136
SARS, 191
Sarver, Calvin, 10, 12, 180–81
Scallops, 122
Science, 155–56
Scientific American, 157
Scimitar cat, 32, 94
Sea level changes, 36–7, 132–33, 136, 142, 193
Sea lions, 139–40
Sea of Cortez, 120–22
Sea otter, 140
Seals, 137, 166
Seasonal mapping, 125
Seri el Desemboque, 121
Seri people, 120–22
Shellfish, 58–9, 120–24, 129, 130–32, 136, 138–39
Shields River (Montana), 10–11, 172, 178
Short-faced bear, 1, 7, 25–7, 49–50, 53–60, 64, 84, 91, 94, 96–7, 103, 115, 126, 141–43, 152, 161, 163, 184, 189, 199
Siberia, 8, 14, 17–8, 24, 38–9, 64–6, 72, 81–2, 107, 109, 128, 174
Siberian tiger, 8, 44–8
Sierra Madre Mountains (Mexico), 48. 108–9
Skeletal remains, 9–18, 20, 75–6, 87, 117, 147, 166–67, 171–72, 174–80
Smithsonian Institution, 182
Snakes, poisonous, 59–60
Solutrean culture, 71–2, 85, 166, 173, 173

Sonora, Mexico, 30–1, 120–22, 167–68

South Carolina, 66, 86, 103, 192

South China Sea, 120

Spain, 71, 73, 85, 120, 166

Spanish influenza, 191

Special Forces, (US Army) 7, 60, 120

Spicer, Edward, 8

Sporomeilla fungus, 194–96

St. Lawrence Valley, 37–8

Stanford, Dennis, 175

Steelhead (trout), 129

Sumatra, 82, 126–27

Syria, 192

Taima-Taima site (Venezuela), 84

Taku River, 136

Tapir, 167, 187

Tasmania, 86

Taylor, D.C., 11–13, 181–82

Tertiary era, 6

Texas A&M University, 155–57, 182

Texas first theory of Clovis, 155–57, 169, 171, 179

Texas, 66, 70, 103, 149, 155–57, 169, 171, 179, 183

Thule people (Inuit), 100, 154–55

Tiburon Island, 120

Tiger, The (Vaillant) 44–8

Tlingit Nation, 127–29, 130, 182

Tompkins, Doug, 45–8, 153–55

Tooth studies, 92, 196–97

Topper, S. Carolina, 151

Trans-oceanic migration route, 71–3, 82–3, 104, 149

Treadwell, Timothy, 110

Udege people (Siberia), 45–6

Umatilla Nation, 76

University of Arizona, 148

University of Calgary, 174

University of Michigan, 6

University of Montana, 11, 181–82

University of Oregon, 159–60, 163

Ural Mountains, 82

Uranium mining, 202

Utah, 103

Vaca, Cabeza de, 18–9

Valentine Creek (Montana), 26–7

Vancouver Expedition, 131

Venezuela, 84

Vietnam War, 7, 60, 120

Virginia, 55, 102–3, 150–51

Walrus, 132

Wanderlust, 165

Washington state, 66, 70, 108, 117, 134, 151–52

Washington Times, 73

Waterfowl, 67, 89, 99, 117, 139, 144, 198

Weapons, of paleoindians, 4–5, 7, 9–10, 12–14, 21, 24, 33, 51, 56–57, 68–69, 72–73, 75, 94, 96, 115, 123, 137, 145, 149–50, 157, 165–69, 172, 175, 188, 191, 198–200

West Arctic Ice Sheet, 36

White supremacy, 76

Whitebark pine, 15, 40

Wilderness, 16, 19–20, 76, 78, 127–28, 202–03

Wilsall, Montana, 147

Wilson, Edward O., 19

Wind River Range (Wyoming), 27, 134

Wisconsin glaciation, 27–8

Wisconsin, 66, 70, 148, 152, 161

Wolf, Gray 32, 60, 88, 97, 108, 111–12, 159, 197

Wolverine, 89, 111

Wood rat, 160

Woolly rhinoceros, 64, 89–90, 91

Wright, Henry, 8

Wyoming, 27, 103, 175

Yana Rhino Horn Site (Siberia), 66–7, 81–2, 87, 91, 93, 97, 100, 115

Yellowstone National Park, 9, 15, 34–5, 49, 64, 76, 110–12, 116, 160, 173,

Yellowstone River, 9–10, 176, 178

Yellowtail, Bill, 182

Younger Dryas (YD) stadial period, 37–8, 165–66, 181–82, 184, 187–88, 192–94, 198–99

Yukon River, 6, 8, 24, 79, 100–2, 184

Zeva River (Siberia), 46

Zoology, 63, 104

Zoonotic diseases, 191–93

AK Press

Ordering Information

AK Press
674-A 23rd Street
Oakland, CA 94612-1163
USA
(510) 208-1700
www.akpress.org
akpress@akpress.org

AK Press
PO Box 12766
Edinburgh, EH8 9YE
Scotland
(0131) 555-5165
www.akuk.com
ak@akedin.demon.co.uk

The addresses above would be delighted to provide you with the latest complete AK catalog, featuring several thousand books, pamphlets, zines, audio products, video products, and stylish apparel published & distributed by AK Press. Alternatively, check out our websites for the complete catalog, latest news and updates, events, and secure ordering.

Also Available from AK Press

The first audio collection from Alexander Cockburn on compact disc.

Beating the Devil
Alexander Cockburn, ISBN 13: 9781902593494 • CD • $14.98

In this collection of recent talks, maverick commentator Alexander Cockburn defiles subjects ranging from Colombia to the American presidency to the Missile Defense System. Whether he's skewering the fallacies of the war on drugs or illuminating the dark crevices of secret government, his erudite and extemporaneous style warms the hearts of even the stodgiest cynics of the left.

Available from CounterPunch/AK Press

Call 1-800-840-3683 or order online from www.counterpunch.org or www.akpress.org

The Case Against Israel
by Michael Neumann

Wielding a buzzsaw of logic, Professor Neumann dismantles plank-by-plank the Zionist rationale for Israel as religious state entitled to trample upon the basic human rights of non-Jews. Along the way, Neumann also offers a passionate amicus brief for the plight of the Palestinian people.

Other Lands Have Dreams: From Baghdad to Pekin Prison
by Kathy Kelly

At a moment when so many despairing peace activists have thrown in the towel, Kathy Kelly, a witness to some of history's worst crimes, never relinquishes hope. Other Lands Have Dreams is literary testimony of the highest order, vividly recording the secret casualties of our era, from the hundreds of thousands of Iraqi children inhumanely denied basic medical care, clean water and food by the US overlords to young mothers sealed inside the sterile dungeons of American prisons in the name of the merciless war on drugs.

Dime's Worth of Difference: Beyond the Lesser of Two Evils
Edited by Alexander Cockburn and Jeffrey St. Clair

Everything you wanted to know about one-party rule in America.

Whiteout: the CIA, Drugs and the Press
by Alexander Cockburn and Jeffrey St. Clair, Verso.

The involvement of the CIA with drug traffickers is a story that has slouched into the limelight every decade or so since the creation of the Agency. In Whiteout, here at last is the full saga.

Been Brown So Long It Looked Like Green to Me: the Politics of Nature
by Jeffrey St. Clair, Common Courage Press.

Covering everything from toxics to electric power plays, St. Clair draws a savage profile of how money and power determine the state of our environment, gives a vivid account of where the environment stands today and what to do about it.

Imperial Crusades: Iraq, Afghanistan and Yugoslavia
by Alexander Cockburn and Jeffrey St. Clair, Verso.

A chronicle of the lies that are now returning each and every day to haunt the deceivers in Washington and London, the secret agendas and the underreported carnage of these wars. We were right and they were wrong, and this book proves the case. Never leave home without it.

Born Under a Bad Sky

By Jeffrey St. Clair

"Movement reporting on a par with Mailer's Armies of the Night"—Peter Linebaugh, author of *Magna Carta Manifesto* and *The Many-Headed Hydra*.

These urgent dispatches are from the frontlines of the war on the Earth. Gird yourself for a visit to a glowing nuclear plant in the backwoods of North Carolina, to the heart of Cancer Alley where chemical companies hide their toxic enterprise behind the dark veil of Homeland Security, and to the world's most contaminated place, the old H-bomb factory at Hanford, which is leaking radioactive poison into the mighty Columbia River.

With unflinching prose, St. Clair confronts the White Death in Iraq, the environmental legacy of a war that will keep on killing decades after the bombing raids have ended. He conjures up the environmental villains of our time, from familiar demons like James Watt and Dick Cheney to more surprising figures, including Supreme Court Justice Stephen Breyer (father of the cancer bond) and the Nobel laureate Al Gore, whose pieties on global warming are sponsored by the nuclear power industry. The mainstream environmental movement doesn't escape indictment. Bloated by grants from big foundations, perched in high-rent office towers, leashed to the neoliberal politics of the Democratic Party, the big green groups have largely acquiesced to the crimes against nature that St. Clair so vividly exposes.

All is not lost. From the wreckage of New Orleans to the imperiled canyons of the Colorado, a new green resistance is taking root. The fate of the grizzly and the ancient forests of Oregon hinge on the courage of these green defenders. This book is also a salute to them.

Yellowstone Drift
Floating the Past in Real Time
By John Holt

High above sea level in the mountains of the Yellowstone National Park plateau, the river tumbles and rushes down to the Paradise Valley just north of Livingston, Montana, before meandering through the northern high plains for well over five hundred serpentine miles to its confluence with the Missouri River in North Dakota. Each chapter of *Yellowstone Drift* chronicles a leg of John Holt's journey down the river, promising that the reader doesn't miss a single mile of natural beauty. Holt, in his customary free-form, anecdotal style and oblique vision, takes the reader on a wild ride down this natural treasure, examining the wildlife, the people, the fishing, and the river itself.

Why We Publish CounterPunch
By Alexander Cockburn and Jeffrey St. Clair

BACK IN 1993, WE FELT unhappy about the state of radical journalism. It didn't have much edge. It didn't have many facts. It was politically timid. It was dull. *CounterPunch* was founded. We wanted it to be the best muckraking newsletter in the country. We wanted it to take aim at the consensus of received wisdom about what can and cannot be reported. We wanted to give our readers a political roadmap they could trust.

A decade later we stand firm on these same beliefs and hopes. We think we've restored honor to muckraking journalism in the tradition of our favorite radical pamphleteers: Edward Abbey, Peter Maurin and Ammon Hennacy, Appeal to Reason, Jacques René Hébert, Tom Paine and John Lilburne.

Every two weeks *CounterPunch* gives you jaw-dropping exposés on: Congress and lobbyists; the environment; labor; the National Security State.

"*CounterPunch* kicks through the floorboards of lies and gets to the foundation of what is really going on in this country", says Michael Ratner, attorney at the Center for Constitutional Rights. "At our house, we fight over who gets to read *CounterPunch* first. Each issue is like spring after a cold, dark winter."

YOU CANNOT MISS ANOTHER ISSUE

Name _____

Address _____

City _____ State _____ Zip _____

Email _____ Phone _____

Credit Card # _____

Exp. Date _____ Signature _____

Type of Subscription: ☐ Gift ☐ Renewal ☐ New Subscriber

Mail check, money order, or credit card info to: CounterPunch P.O. Box 228 Petrolia, CA 95558. All renewals outside the U.S. please add shipping: $20.00 per year for postage, for Canada and Mexico; all other countries outside the US add $30.00 per year. The information you submit is confidential and is never shared or sold. Please give us your phone number, so that we may contact you in case of any questions with your renewal, or if there is ever a problem with your subscription.

☐ 1 year, print **$55**
☐ 1 year, email **$35**
☐ 1 year, both **$65**
☐ 1 year, reduced* **$45**
☐ Supporter **$65**
   1 Year, Either

Mail Renewals to: P.O. Box 228 Petrolia, CA 95558
1 (800) 840-3683 **or** 1 (707) 629-3683 www.counterpunch.org